**Springer Textbooks in Earth Sciences,
Geography and Environment**

The Springer Textbooks series publishes a broad portfolio of textbooks on Earth Sciences, Geography and Environmental Science. Springer textbooks provide comprehensive introductions as well as in-depth knowledge for advanced studies. A clear, reader-friendly layout and features such as end-of-chapter summaries, work examples, exercises, and glossaries help the reader to access the subject. Springer textbooks are essential for students, researchers and applied scientists.

Jennifer Pontius · Alan McIntosh

Environmental Problem Solving in an Age of Climate Change

Volume One: Basic Tools and Techniques

 Springer

Jennifer Pontius
University of Vermont
Burlington, VT, USA

Alan McIntosh
University of Vermont
Burlington, VT, USA

ISSN 2510-1307 ISSN 2510-1315 (electronic)
Springer Textbooks in Earth Sciences, Geography and Environment
ISBN 978-3-031-48761-3 ISBN 978-3-031-48762-0 (eBook)
https://doi.org/10.1007/978-3-031-48762-0

This Springer imprint is published by the registered company Springer Nature Switzerland AG
The registered company address is: Gewerbestrasse 11, 6330 Cham, Switzerland

Paper in this product is recyclable.

This book is dedicated to the students of the University of Vermont's Rubenstein School of Environment and Natural Resources who have inspired us with their passion and energy and their commitment to tackling today's most pressing environmental issues. You give us hope for the future of our planet.

Contents

1.1 The Far-Reaching Impacts of Climate Change

Climate change is the most pressing environmental issue of our time. The 2022 report from the Intergovernmental Panel on Climate Change documents how human-induced climate change is causing widespread impacts on ecosystems and affecting millions of people around the world. Heatwaves, floods, and droughts have already exceeded the adaptive memory of many ecosystems, driving mass mortalities among species of trees, mammals, and fish (IPCC 2022; Tye et al. 2022). Millions of people are regularly exposed to extreme weather events, degrading infrastructure and increasing food and water insecurity.

The IPCC report also projects how our climate will continue to change. Their forecast includes warnings of unavoidable climate hazards over the next two decades, with increased risks to human populations. In addition to direct impacts from changing temperature and precipitation patterns, climate change indirectly influences ecosystem processes. This means that climate change has the potential to worsen a host of local environmental problems. Conversely, other seemingly unrelated environmental issues also can directly contribute to ongoing climate change. This puts climate change at the center of almost all environmental issues, from local to global scales, including terrestrial, marine, and societal challenges (Fig. 1.1).

While the climate crisis may seem insurmountable, society does have the power to tackle the challenge in several important ways:

1. Taking ambitious, accelerated actions to **reduce our greenhouse gas emissions** will help to slow climate change and avoid the most dire projections.
2. Tapping into ecosystems' natural ability to **capture and sequester carbon** can help offset some of our emissions.
3. Helping both ecological and social systems adapt **to climate change** will minimize impacts.

4. Building the realities of climate change into all the current work we do will help safeguard **and strengthen nature while addressing local environmental issues**.

This book is designed to help you build your capacity in that last arena: **addressing important local environmental issues while considering the impacts of climate change on potential solutions**. This is at the heart of what environmental scientists do: identify, understand, manage, and solve today's pressing environmental issues. Typically, this work is accomplished at local to regional scales, with a focus on unique problems with specific desired outcomes. While issues such as salmon conservation, waste management, or deforestation may seem far removed from climate change, for any solutions to be truly sustainable, current climate impacts and future climate threats must be considered.

1.2 Tackling Environmental Challenges

Addressing complex environmental issues requires a working knowledge of both basic and applied science, including disciplinary frameworks for understanding complex, interconnected systems and decision support to help identify the "best" approach to solve the issue. Truly sustainable solutions must also incorporate social and economic concerns like environmental justice, social equity, and economic viability. Add to this the ongoing impacts of climate change and uncertainty around climate projections, and it becomes clear that tackling environmental challenges usually requires a team of environmental professionals trained and practiced at tackling complex problems.

In this text, you're able to practice solving complex environmental problems with a selection of 12 current environmental issues (Fig. 1.2). From desertification in sub-Saharan Africa to Dead Zones in the Gulf of Mexico, what may seem like localized environmental issues actually result from a

J. Pontius, A. McIntosh, *Environmental Problem Solving in an Age of Climate Change*, Springer Textbooks in Earth Sciences, Geography and Environment, https://doi.org/10.1007/978-3-031-48762-0_1

Fig. 1.1 The IPCC report has a strong focus on the interactions among the coupled systems climate, ecosystems, and human society. These interactions form the basis of emerging risks from climate change, ecosystem degradation, and biodiversity loss and, at the same time, offer opportunities to better manage climate change in the future. (Source: IPCC (2022) {Public Domain})

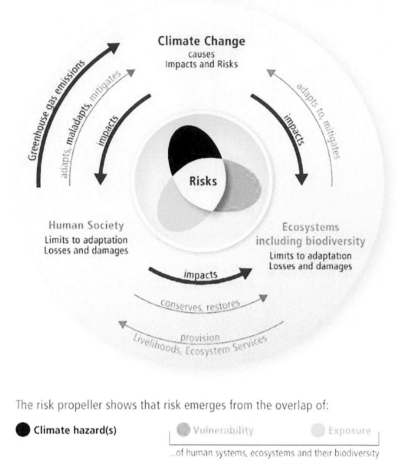

variety of human activities connected to global-scale processes like climate change.

In addition to helping you apply your basic understanding of ecological processes, mapping the complexity of the larger socioecological system, and researching and comparing solutions, this text is designed to help you see how environmental professionals apply their scientific knowledge in a team setting to tackle complex problems.

The opportunity to tackle grand environmental problems is both rewarding and challenging. The rewarding part comes from knowing that your efforts will help create a more sustainable future. The challenging part is the enormity of complex global issues like climate change. The key to successful environmental work is focusing on issues where we can affect change. Looking for a solution to "climate change" can be daunting, but identifying an approach for mitigating the impacts of climate change in a smaller system is achievable. It is through collective work on these more tangible issues that environmental professionals are ensuring a brighter future for nature and the people who depend on it.

1.3 User's Guide

This text is designed to help prepare the next generation of transdisciplinary environmental professionals. Rather than focus on only the natural science or social science aspects of these issues, this text stresses the importance of bridging the gap among disciplines. Sustainable solutions to complex environmental problems must consider the biological, chemical, and physical components of the ecological system, but they must also factor in the influence and impacts of humans as a part of this system. How humans contribute to, are impacted by, and can work towards solving environmental problems are central to the work environmental professionals do. Given this reality, you need to consider stakeholder perspectives, economic considerations, cultural values, governance, and policy implications as a part of the process.

All 12 units in this book use the same basic structure, employing the same tools and frameworks to identify solutions to complex environmental issues. Exposure to, and practice with, these disciplinary frameworks will prepare

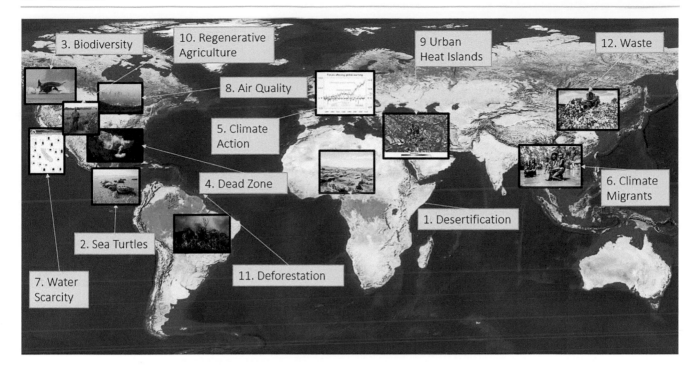

Fig. 1.2 This text provides project-based environmental problem-solving challenges from around the globe

you to apply your understanding of natural science within the context of complex socioecological systems. While you will work to understand the problems presented, the focus will be on working towards finding a sustainable solution.

It is important to understand that what is presented in each of these units is part of a much more complex issue. There may be additional factors involved that we may not discuss in the text. Be prepared to apply your own understanding of these systems to better inform your choice of a solution. Similarly, there may be other possible solutions that could be considered. You will have a chance to reflect on such alternatives as you work through each unit, including an opportunity to flesh out and propose your own alternate solution versus any of the three solutions we present in the text.

Each unit focuses on a different environmental issue with a specific challenge for you to address. Working with your classmates, you will build your foundational knowledge around this issue before exploring and comparing three possible solutions to address your challenge. Your final objective is to explain and justify your chosen solution using clear science communication techniques.

Unit Components Each of the 12 units in the book is comprised of the following elements:

- **Background** information to build your foundational knowledge about the focus issue, including how it is related to climate change. Throughout this section, you

will find links to additional resources to explore important topics.

- An introduction to your specific **Unit Challenge**, including the three options you'll consider as solutions and the relevant facts and assumptions you'll need to consider when choosing a solution.
- **Discovery activities**, which require you to explore the scientific literature and share the background information you'll need to make an informed decision.
- **Analysis exercises,** where you work at rotating stations on data sets to examine the evidence around each of your potential solutions.
- **Solution evaluations**, during which you compare the pros and cons of each of your three options and solicit input from stakeholders before you make your final choice. Note that while we provide three practical solutions for each challenge, you may choose to flesh out and propose a solution of your own as well.

Each unit identifies a number of key disciplinary and professional skills that you'll need to use in order to complete the required unit exercises (e.g., Navigating Scientific Literature; Decision Support, Scientific Communication, etc.). This text is designed to allow you to practice and apply these to real-world issues. However, if you need to review these skills, a good reference for an overview of common frameworks for key disciplinary skills can be found in our Springer textbook: **Critical Skills for Environmental Professionals (Pontius and McIntosh, 2020)**. Your final task in each unit will be to apply these skills to identify, justify and summarize your recommendations to solve the environmental issue and address your Unit Challenge in a concise **Fact Sheet** that can be broadly shared with a diverse group of key stakeholders.

Throughout the book, you'll see links connecting you to a variety of information sources. Some of these are from the peer-reviewed literature, but many are from the gray literature, including state, federal, and international agency and Non-Governmental Organization (NGO) reports. While you should always treat the gray literature carefully and be wary of such issues as bias, these sources can often yield useful information as you tackle your Unit Challenges.

Professionals rarely tackle and solve environmental issues on their own; they're almost always part of a team. Not only are many different disciplinary backgrounds needed to find the best solution but having a broad sounding board helps improve the chances that innovative approaches will be considered and the best option will ultimately be chosen. While tackling the Unit Challenges, you'll have ample opportunity to interact with others on this issue.

Ultimately, working on project-based climate-impacted challenges like these will allow you to build your skills as an environmental professional. Through these examples, you will see that climate change and its impacts on the natural world and human populations can be tackled. The key is to work collectively and collaboratively on tangible issues we can directly address. In aggregate, these efforts can "move the needle" on climate change and help to build a brighter, more sustainable future.

References

IPCC (2022) Summary for policymakers. Pörtner HO, Roberts DC, Poloczanska ES, Mintenbeck K, Tignor M, Alegría A, Craig M, Langsdorf S, Löschke S, Möller V, Okem A (eds). In: Climate change 2022: impacts, adaptation and vulnerability. Contribution of Working Group II to the Sixth Assessment Report of the Intergovernmental Panel on Climate Change. Pörtner HO, Roberts DC, Tignor M, Poloczanska ES, Mintenbeck K, Alegría A, Craig M, Langsdorf S, Löschke S, Möller V, Okem A, Rama B (eds) Cambridge University Press, Cambridge/New York, pp 3–33

Pontius J, McIntosh A (2020) Critical skills for environmental professionals: putting knowledge into practice. https://doi.org/10.1007/978-3-030-28542-5.1

Tye SP, Siepielski AM, Bray A, Rypel AL, Phelps NBD, Fey SB (2022) Climate warming amplifies the frequency of fish mass mortality events across north temperate lakes. L&O Letters. https://doi.org/10.1002/1012.10274

Core Knowledge

Soils, Agriculture, Land use, Human population, Food insecurity

2.1 Environmental Issue

Desertification, the process by which fertile lands become desert-like, typically results from some combination of drought, deforestation, and unsustainable agricultural practices. It occurs primarily in arid, semiarid, and dry areas collectively known as drylands. Given that drylands cover nearly half of the planet's land area and are home to roughly 3 billion people, the threat of desertification is a serious one.

Because desertification reduces the amount of land that can support crops and livestock (Fig. 2.1), it is directly linked to issues of food insecurity. Desertification occurs across the globe, including areas in Europe and the United States, but the problem is particularly severe in Asia and Africa, where climate change and large human populations make this issue a vital concern.

The food insecurity issue in Africa is complicated, with many factors contributing to the problem. In this unit, you'll explore the causes of desertification in sub-Saharan Africa, how it is linked to food insecurity, and why large-scale, coordinated solutions are necessary as the climate changes.

2.2 Background Information

2.2.1 The Problem

Desertification is not just about drought conditions. In ecosystems already adapted to low rainfall, it is instead a slow process driven by soil degradation that begins with the disappearance of stabilizing vegetation, followed by the loss of topsoil from wind and water erosion. This results in a dramatic reduction in the ability of the landscape to retain the water it does receive and to support agriculture. The forces causing desertification are both natural (e.g., long-term changes in rainfall patterns that limit the growth of stabilizing vegetation) and human (e.g., overgrazing, deforestation, and other unsustainable agricultural practices). While this process has occurred throughout human history (e.g., the 1930s Dust Bowl in the United States), changing rainfall patterns in semiarid regions have greatly accelerated the rate and expansion of desertification across the planet. Figure 2.2 highlights the complex linkages and feedbacks among climate change, land use, soil health, ecosystem processes, and biodiversity

Currently, desertification is a particular concern in Africa (Fig. 2.3), where prolonged droughts have left millions in a state of food scarcity The World Hunger Education Service (WHES) (2018) estimated that in 2016, while 9.3% of the world's population suffered from severe food insecurity, the figure in Africa was 27.4%, with about 333 million people affected. WHES noted that areas in eastern Africa were particularly hard hit, with an estimated one-third of the population undernourished.

A cruel cycle develops. As more people need to cultivate food on marginal dry landscapes, the process of desertification accelerates, reducing the amount of arable land even further. Already limited in their ability to attain food security, many Africans now face the prospect of continuing losses of arable land. As more and more drylands can no longer support any form of crop or livestock production, the extent of food insecurity will only increase.

2.2.2 The Role of Climate Change

Intensification of droughts in semiarid regions where physical systems are already vulnerable to degradation serves as a catalyst for desertification to accelerate. An important contributor to the process is water scarcity, either because of low

J. Pontius, A. McIntosh, *Environmental Problem Solving in an Age of Climate Change*, Springer Textbooks in Earth Sciences, Geography and Environment, https://doi.org/10.1007/978-3-031-48762-0_2

Fig. 2.1 Antidesertification
sand fences south of the town
of Erfoud, Morocco. (Source:
Anderson Sady [CC 3.0] via
Wikimedia Commons)

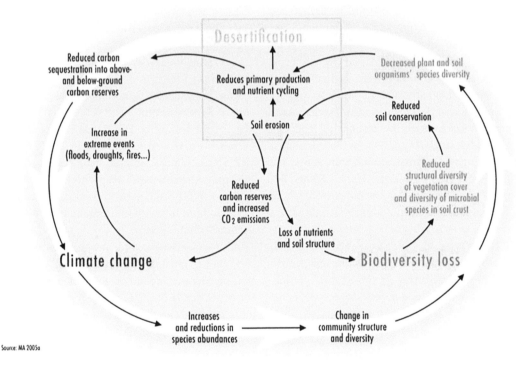

Fig. 2.2 Linkages and feedback loops among desertification, global climate change, and biodiversity loss. (Source: GRID-Arendal, via Flickr
{Public Domain})

precipitation and water availability or greater water demand
for irrigation and urban and industrial uses.

The climate has already been changing rapidly in Africa,
with projections for continued warming and shifts in precipitation patterns (Blunden and Arndt 2020). According to the
Intergovernmental Panel on Climate Change (IPCC) (2014),
the frequency, intensity, and duration of droughts are

expected to increase around the globe, indicating that desertification will likely continue to be a problem of great
concern.

Direct impacts on crops and livestock will occur, while
indirect effects will include increased soil erosion and
greater pest and disease damage to crops and livestock
caused by higher temperatures. As climate disruption con-

Fig. 2.3 Desertification vulnerability of Africa. (Source: USDA Natural Resource Conservation Service {Public Domain})

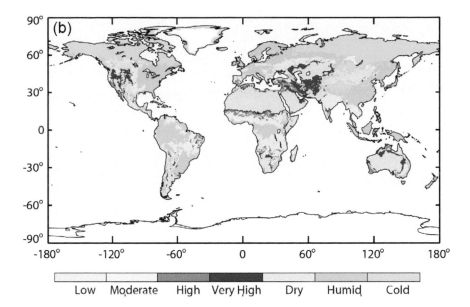

tinues, mass migration within Africa will likely worsen food insecurity, as nations which have sufficient food supplies for their populations will face pressures to provide food to those in need.

2.2.3 Solutions

There are many ways to tackle the issue of desertification, from preventing its expansion to restoring lands already impacted. Grand efforts to prevent the spread of existing deserts have been initiated across Africa. For example, the Great Green Wall was envisioned as an 8000 km long swath of trees crossing the entire continent to halt the movement of Saharan sands. However, most of the planted trees did not survive in the harsh climate. Instead, the African Union found that indigenous land use techniques were most successful at halting the advance of deserts. Rather than rely only on a strip of trees, sustainable land use practices can create a mosaic of stable vegetation to keep soils from eroding and degrading (Fig. 2.4).

Other agricultural techniques that farmers and other landowners can use to improve and revive drylands (IPCC 2019) include adding clay or organic material to soils to improve drainage and increase their ability to hold water. Practicing crop rotation; adding soil conditioners like lime, gypsum, and sulfates; controlling the intensity and duration of grazing; and sowing new varieties of grass seeds are additional steps that can help. The challenge, of course, is having sufficient resources to make such tools available to farmers in critical areas and to educate farmers to help ensure the widespread use of such tools.

Restoration and rehabilitation of lands already degraded are also possible. These processes work to restore ecosystem structure and services to previous conditions. Approaches may include planting specific species, preventing erosion with physical structures, and enriching soils with nutrients.

While these are important steps that can be applied at the individual landowner level, given the scope and extent of the threat posed to the vast expanse of drylands by climate change, it seems prudent to also focus on broader efforts in the sub-Sahara. Tackling something this pervasive will require cooperation among the affected nations.

2.2.4 Unit Challenge

If the worst impacts of desertification in sub-Saharan Africa are to be avoided, all parties, including national leaders, scientists, and NGOs, will need to work together to change land use practices across the region. Your Unit Challenge is **to help local farmers combat desertification**. The exercises that follow in this unit will help you identify the best approach or combination of approaches to attack the problem.

2.2.5 The Scenario

As an agricultural expert and field scientist at the United Nations' Food and Agricultural Organization (FAO), you are working with a team of experts to evaluate several approaches that individual landowners can use to fight further desertification in sub-Saharan Africa. The aim is to identify which approach could be most widely applied by local communities throughout sub-Saharan Africa to protect drylands and prevent further desertification while meeting peoples' basic needs for food, cooking fuels, income stability, and clean water.

Fig. 2.4 An aerial view of agroforestry management practices in Niger in 2004. (Source: USGS {Public Domain})

International publications from the IPCC, the UN's FAO, and the World Resources Institute (WRI) have identified a host of possible approaches that communities might use to try to slow and reverse desertification on their lands.

After a detailed review of available approaches, your group decides to consider three different techniques:

1. **Planting ground cover species**. Establishing surface-rooting vascular plants that require comparatively little water to survive can promote soil stabilization, increase soil nutrient levels, and enhance levels of soil organic matter (SOM). Some species providing ground cover may also have commercial value and represent a source of income for local communities.

    Specific recommendation: Disseminate *Jatropha* seeds for widespread plantings. Thriving in arid regions, many of the 170 species of the genus *Jatropha* (Fig. 2.5) are resistant to pests and can improve soil quality because the plants are rich in nitrogen, phosphorus, and potassium. *Jatropha curcas* beans are also used as a source of biodiesel fuel.

2. **Supporting local farmers**. With climate change increasing the risk of erosion, available sustainable agricultural practices, including mulching with crop residues to enhance soil organic carbon levels, practicing diversified crop rotation, and adopting mixed use in which a single farm combines livestock rearing and cropping to promote nutrient recycling in soils, will be considered. Such practices not only add to the SOM pool and increase crop yield, but, by sequestering carbon, they can help offset carbon emissions from fossil fuels.

    Specific recommendation: Support local conservation resource managers in their efforts to educate local communities about sustainable farming techniques and provide incentives to adopt soil conservation strategies.

3. **Adopting innovative water management techniques.** With more persistent and severe droughts making desertification worse, successful water management is a crucial step in fighting the problem. Available approaches include the use of traditional water harvesting techniques, water storage, water capture during intensive rainfall, improving groundwater recharge, and floodwater spreading.

    Specific recommendation: Provide financial resources and expertise to build water management structures. In very dry climates, ancient approaches for managing scarce water resources can be effective in promoting water conservation and maintaining soil moisture. Stones bunds, which are walls of stones placed along slopes, slow water runoff, and increase soil infiltration. Similarly, zais, pits dug to catch water and concentrate compost, increase water availability in drylands, and promote soil fertility.

Each of these three approaches has important advantages and disadvantages and costs and benefits. Your task is to evaluate each and recommend the approach that will, in your opinion, perform best at slowing desertification in drylands across sub-Saharan Africa as the climate continues to change.

Fig. 2.5 *Jatropha aspidoscelis*. (Source: Patrick Alexander, (CC BY 2.0) via Wikimedia Commons)

2.2.6   Relevant Facts and Assumptions

- Your goal is to be able to manage 10,000 ha using one of these approaches.
- Each *Jatropha* seedling costs 3 rupees. Planting density is about 200 seedlings per hectare (ha). In subsequent years, the annual labor costs (e.g., planting, weeding, etc.) are about 100 rupees per ha.
- Local residents can be trained and employed as Natural Resource educators for about 50,000 rupees/year and can be placed at 50 population centers across the region, with each center able to manage 200 ha. Additional incentives and materials, including seeds and planting costs for cover crops that serve to replenish SOM, are estimated to total 100 rupees/ha/year. This would be a yearly, recurring cost to operate the program.
- Local residents can be trained and employed to build zais, stone bunds, and planting pits. This would require an initial investment of 400 rupees/ha and in subsequent years a recurring annual maintenance and repair cost of 100 rupees/ha.
- Assume that all three options could be equally effective at combating desertification at the locations where they are implemented. Your goal is to maximize implementation across the region, while providing local communities with solutions that are popular and likely to be maintained long-term.
- The soils in the area of concern are mostly Arenosols, sandy soils with very little organic matter supporting limited amounts of specialized vegetation. They are low in humus and lack subsurface clays.

- The climate in the region is dry tropical, with a short rainy season and a long dry season; rainfall is about 50 cm annually
- Local populations depend on farming and grazing for subsistence and income.

2.2.7   Build Your Foundational Knowledge

Below are web sources that provide additional information about each of the solutions you're considering for this Unit Challenge. This information can build a critical foundation to help you evaluate each option and support your final choice. After reviewing each source, be prepared to answer questions in the Preparation Assessment Quiz and to summarize any information relevant to your Unit Challenge.

Planting ground cover:
Jatropha curcas: An environmental silver bullet,
Jatropha curcas as a multipurpose crop

Supporting local farmers:
Sustainable agriculture will help stop desertification,
Millennials are transforming African farming

Adopting innovative water management techniques:
Water conservation slows desertification,
Soil and water conservation in Burkina Faso, West Africa

Final Product: A one-page Fact Sheet summarizing the issue, detailing your solution, and justifying your choice. Consider your audience, decision-makers at the World Bank who will use this Fact Sheet to determine the best actions in which they should invest critical support. Be sure to demonstrate how your proposed solution will stand up to the challenges posed by climate change.

2.2.8 Preparation Assessment Quiz

Are you ready to tackle your challenge? At this point, you should understand the basic environmental principles and ecological processes involved in this environmental problem. Consider the following questions. If you are comfortable with answering these, then you are ready to head into Discovery, Analysis, and Solutions activities.

- What is the difference between drylands and desertification?
- What are the primary drivers of desertification?
- What is food insecurity?
- How does desertification contribute to food insecurity?
- How will climate change likely worsen the severity of food insecurity in sub-Saharan Africa?
- The Borgen Magazine article on *Jatropha* claims that 1 ha of this shrub can capture how much CO_2 annually?
- According to the World Economic Forum article on millennials in Africa, what is an "agripreneur"?
- According to Nyamekye et al., how many zais/ha are there on an average farm?
- For each of your proposed solutions, are there any additional benefits that might arise from their implementation that might not be directly related to your Unit Challenge?
- For each of your proposed solutions, are there any negative unintended consequences that might result from their implementation?
- What additional information did you glean from your web sources that might help inform your Unit Challenge?

2.3 Desertification: Discovery

Specific Skills You'll Need to Review: Navigating the Scientific Literature, Science Communication, Problem-Solving

2.3.1 Independent Research

(Key Skill: Navigating the Scientific Literature)

To better understand the various approaches being considered to reverse the process of desertification in drylands, you first need to examine the literature to see what others have found. Conduct a search of the peer-reviewed scientific literature focused on the solution you have been assigned and identify one research paper that focuses on your assigned approach.

Prepare a summary of the article you selected that includes the following:

- **Citation**
- **Main topic**: Stick to a few words, likely pulled from the title.
- **General summary**: A few bulleted sentences summarizing the research question it addresses and approaches it takes.
- **Methods:** How did they approach their research question?
- **Location**: Where was the work done?
- **Conclusions**: Concise list of the findings, specifically capturing the take-home message.
- **Relevance**: How might this study help inform your Unit Challenge? Feel free to make a bulleted list of information you may want to include later.

2.3.2 Literature Share: Reciprocal Instruction

(Key Skill: Scientific Communication)

Share In small groups, share and critique the research article you found. Keep in mind that your peers have not read this article, and it is your job to convey the key information to them. Note the items that will be important to consider when you are developing your solution to the challenge.

Critique Evaluate how these studies might help inform your Unit Challenge. Consider the following:

- Source (Quality of the work or bias of the authors)
- Methods (Did their methods sufficiently address the research question?)
- Conclusions (Did the results justify the conclusions made?)
- Relevance (Can these findings be applied to your challenge?)

Based on your critique, choose one article to share with the larger class, along with the key information that may be useful in deciding on a solution to propose.

2.3.3 Think-Compare-Share

(Key Skill: Problem-Solving)

Now that you have more information about possible solutions for this unit's challenge, you need to **develop a more formal problem definition** to guide your work throughout the rest of the exercises.

Think Start by working independently to develop a specific Problem Statement to guide the remainder of your work. Problem Statements provide the relevant information and boundaries to make the issue something you can effectively assess and tackle. The basics of a formal Problem Statement include the following:

Problem Statement: A short, concise statement summarizing the issue that includes the following:

- A **description** of the undesired condition or change that you hope to achieve (What is the actual problem?)
- **Justification** for addressing the problem (Why does this problem matter?)
- Potential **sources** or **causes** of this problem (What is the cause you need to address?)
- The **metrics** you will use to assess the status of the problem (How will you know if you are making a difference in the problem?)
- The **desired outcome** for these metrics (What is the end goal or ideal state?)
- Potential **solutions** to consider (How might you attempt to achieve this goal?)

Compare/Share Now return to your small group to share your Problem Statements. Use each of your ideas to develop a joint statement that contains all key information and is concise, clear, and well written.

Unit Discovery Summary *Submit a final Problem Statement that succinctly captures the key information to guide your work on this Unit Challenge.*

2.3.4 Reflecting on Your Work

(Key Skill: Personal Reflection)

After your work in Discovery, you should have a better idea of the problems you face and have produced a Problem Statement you can use to tackle the Unit Challenge. Take a moment to reflect on this work.

Consider the following prompts, but feel free to expand on any to best capture your learning experience and better inform your next steps.

- Of the skills you practiced in Discovery, which were the most challenging? Which were the most interesting?
- How were you most comfortable working during these exercises? In small groups, independently, or with the larger class? Why? How does your choice reflect your personality type and leadership style?
- Was your Problem Statement strictly focused on the environmental problem, or did it also consider important social and economic considerations? How might a focus on the environmental aspects limit your ability to identify truly sustainable solutions?
- You've been given three viable solutions to assess as a part of this case study. However, this is not an exhaustive list of options or even necessarily the best possible course of action for every scenario. Take a moment to "think outside the box." Are there any other possible solutions you think would be worth exploring? Describe one that you think would be worth pursuing.

2.4 Desertification: Analysis

Specific Skills You'll Need to Review: Quantitative Literacy, Sustainability Science

Review your Background and Discovery sections before beginning the Rotating Station exercises below. While you focused on one potential solution in your Independent Research in Discovery, keep an open mind as your work through Analysis activities.

2.4.1 Rotating Stations

(Key Skill: Quantitative Literacy)

At each of the following stations, you will review data that are relevant to the three potential solutions you're considering. Spend some time working through the analyses at each station to learn more about this issue and possible solutions for your Unit Challenge.
Be sure to write down one finding at each station that will help inform your selection of a solution.

Station 1: Planting Ground Cover While *Jatropha* plants have the potential to stabilize soil, replenish nutrients, and provide alternative economic benefits from the production of biofuels, it isn't suited to grow everywhere. Tedesse (2014) developed a habitat suitability model for *Jatropha curcas*

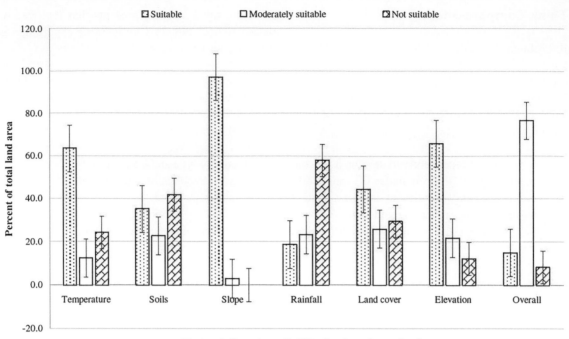

Fig. 2.6 Suitability analysis for *Jatropha* production based on various landscape characteristics as a percent of land area suitable, moderately suitable, and not suitable. Left bar = suitable; middle bar = moderately suitable; right bar = not suitable. (Source: Tedesse (2014) {Open Access})

across various regions of Ethiopia (Fig. 2.6). Results include the following:

Based on the results of this study, answer the following questions:

- Which habitat characteristic is most restrictive (where is planting not suitable across the largest area?)?
- What are the characteristics of locations where *Jatropha* plantings would fare the best?
- Assume that *Jatropha* prefers cooler temperatures, level slopes, and higher rainfall; how would this figure change with projected increased temperatures and decreased rainfall?
- Overall, what proportion of land is suitable or moderately suitable for *Jatropha* planting?
- Assuming that the region contains 100,000 ha, about how many ha would be suitable for *Jatropha* plantings?

Record any relevant Station 1 findings

Station 2: Supporting Local Farmers A study of factors influencing the adoption of ecosystem-based farm management practices in Ghana analyzed how likely farmers were to adopt eight different sustainable farming practices with or without support from extension officers

Table 2.1 Coefficient estimates for factors that influence adoption of sustainable agricultural practices (note that * indicates significant influence on adoption where * = 0.10 significance, ** = 0.05 significance, and *** = 0.01 significance)

| Variables | Estimates of negative binomial model (NBM) | | | |
| | CIS | | GIS | |
	Coeff.	SE	Coeff.	SE
Constant	1.016	0.296	0.044	0.318
Age	0.002	0.004	0.011	0.005**
Sex	0.016	0.085	−0.022	0.132
Educ_d	0.040	0.091	−0.025	0.105
Ext_visits_d	0.160	0.094*	0.048	0.102
Fm_dist. (km)	−0.234	0.083***	−0.103	0.055*
Fm_size (acres)	0.040	0.106	0.005	0.046
Soil_perception	0.159	0.089*	0.239	0.123*
Knw_EBFMPs	0.016	0.011	0.040	0.015***

Positive coefficients indicate a direct relationship between variables and the likelihood of adoption, while negative values indicate an indirect relationship with less likelihood of adoption. Source: Agula et al. (2018) {Open Access}

promoting community-managed (CIS) instead of government-managed (GIS) landscapes (Table 2.1). Modeled variables included farmer age, sex, level of education (Educ_d), extension visits (Ext_visits_d), irrigation distance (Fm_dist), farm size for irrigation (Fm_size), perception of farm fertility value (Soil_perception), and farmer knowledge of ecosystem-based farm management practices (Knw_EBFMPs).

- Note that for both CIS and GIS, * indicates various levels of significance for each variable. Based on these results, which factors are significant in determining whether or not sustainable agricultural practices are adopted?
- Describe the nature of those relationships (direct or indirect based on the sign of the coefficient) and how adoption could be maximized in both CIS- and GIS-managed farmlands.
- Based on these variables, where would extension work be most successful?

Record any relevant Station 2 findings

Station 3: Adopting Innovative Water Management Techniques In many locations, the loss of water through runoff, soil evaporation, and drainage below the root zone is more responsible for water deficiencies than inadequate rainfall. The use of pits, or zais, and stone bunds has been shown to be an effective way to increase water conservation and effectively rehabilitate previously abandoned and degraded bare lands in sub-Saharan Africa.

The data below (Table 2.2) compare the potential costs and revenues of various water conservation strategies in West Africa. This includes a control following current agricultural practices; soil scarifying to allow for better infiltration of water; manually dug zais to capture and hold water and organic matter; larger mechanized zais with additional excavation to increase water holding capacity; and larger mechanized zais built without additional excavation.

- Of the various management techniques, which has the lowest short-term total cost?
- Which has the greatest income potential?
- Are any of these worth implementing (overall benefit outweighs the costs)? Which approach would you recommend?
- What other factors besides the simple economic cost/benefits should be considered?

Record any relevant Station 3 findings

Station 4 For your Unit Challenge, you were given a set of assumptions that included estimated costs for implementing each of your three possible solutions. If the goal is to implement your chosen solution across 10,000 ha of drylands in sub-Saharan Africa, estimate the following:

- What are the initial implementation costs (Year 1) for each possible solution?
- What are the 5-year implementation plus maintenance costs for each possible solution (Year 1 startup costs plus four more years to continue the programs)?
- Is there a considerable difference in the costs for each of these? How important are startup costs vs. the longer-term maintenance costs? How much should these costs weigh in the selection of any one potential solution?

Record any relevant Station 4 findings

2.4.2 System Mapping

(Key Skill: Sustainability Science)

Your assigned solution may have a variety of direct and indirect economic, social, and ecological impacts that should also be considered. For example, planting a cover crop like *Jatropha* may provide additional economic benefits to a farmer, while building and maintaining a system of stone walls or bunds would require substantial hard labor.

Working with other class members assigned the same solution, develop a simple sustainability map that shows the various system connections across each of the three sustainability domains (ecological, economic, and societal). When you find connections between impacts that cross domains, draw a line to identify the connection (e.g., production of a commercial by-product (economic) leading to increased financial resources for a family (social)).

Table 2.2 Economic evaluation of the treatments at Pougyango (in $CFA/ha where CFA is the West African Franc)

	Control	Soil scarifying	Manual *zaï*	Mechanized *zaï* with excavation	Mechanized *zaï*
Soil tillage cost	0	3,000	55,000	26,900	15,000
Total cost	58,333	69,003	166,698	93,659	76,790
Income grains	29,820	49,560	146,160	201,740	191,380
Income straw	12,668	14,220	37,808	56,970	52,748
Total income	42,488	63,780	183,968	258,710	244,128
Benefit (income- cost)	−15,846	−5223	17,270	165,051	167,337
Additional costs	–	0	52,000	23,900	12,000
Total benefits	–	0	120,188	194,930	180,348
Ratio cost/benefit	–	–	2.31	8.16	15.03

Source: Zougmore et al. (2014) {Open Access}

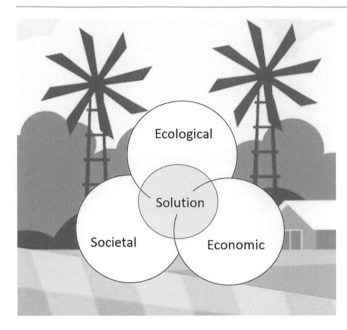

Fig. 2.7 Template sustainability map to help connect ecological, societal, and economic considerations associated with your potential solutions

The goal is to think broadly about this larger system and envision how implementing actions in one domain (e.g., ecological) may impact components in another domain (e.g., economic or societal).

Use the following template (Fig. 2.7) to get started:

When each group has completed their basic sustainability map, come together as a class to compare maps for the three possible solutions. This information will help inform your decision support work later in the unit.

Unit Analysis Summary Based on your explorations, what have you learned that can help inform your choice of a solution? Do the data support the adoption of one of these potential solutions?

2.4.3 Reflecting on Your Work

(Key Skill: Personal Reflection)

During your work in Analysis, you explored some of the research into possible solutions to help inform your decision. Take a moment to reflect on this work. Consider the following prompts, but feel free to expand on any to best capture your learning experience and better inform your next steps.

- How did you feel working with data? Do you consider quantitative literacy a strength or an area for improvement for you?

- How important should science be in informing management and policy? Do you feel the data you examined support and justify the costs of addressing your challenge?
- Any solution should be examined using a sustainability science lens. Your goal is to work with farmers to slow the process of desertification in sub-Saharan Africa. However, any solution you implement may have other direct and indirect impacts. What are other possible economic, social, or ecological impacts? How should these considerations influence your decision?

2.5   Desertification: Solutions

Specific Skills You'll Need to Review: Problem-Solving, Decision Support, Communicating Science

Review your Unit Challenge and the major findings from Discovery and Analysis, including the sustainability map you created to highlight connections among possible solutions and the larger socioecological system.

2.5.1   Small Group Guided Worksheets

(Key Skill: Decision Support)

Decision support matrices can help break down the desired outcomes to reflect multiple criteria for consideration, and they allow you to compare how each possible solution achieves those desired outcomes. This not only helps inform decision-making, but it also provides transparency in the decision process and justification to help you advocate for adopting a particular solution.

For the three possible solutions, you will evaluate how well each can achieve the following desired outcomes:

- Helps slow the process of desertification by stabilizing vulnerable landscapes
- Supports local communities and economies
- Keeps 5-year implementation and maintenance costs to a minimum
- Provides a lasting impact (not a short-term fix), even with a changing climate
- Is widely acceptable to stakeholder groups
- Minimizes secondary impacts (e.g., to other animal species, local economies)

Considering your three potential solutions, develop a formal decision support matrix to compare and evaluate each approach using the template matrix below (Table 2.3). Getting the most out of the decision matrix requires a depth of knowledge about each of the three possible solutions.

Table 2.3 Basic decision matrix to compare the three possible solutions for your Unit Challenge. Include your justification for each solution's ranking for each desired outcome

Solution Options	Desired Outcomes					
	Effective at reducing desertification	Supports local communities / economies	Minimizes costs	Maximizes longevity of impact	Maximizes stakeholder buy-In	Minimizes secondary impacts
Planting *Jatropha*						
Sustainable agriculture education / extension						
Water management (bunds and zais)						
Justification for Ranking						

Use a scale of 1-5 to score each option under each desired outcome category where 1 = does not achieve the desired outcome and 5 = completely achieves the desired outcome.

Below we list additional sources about each option. Please review each of these, paying particular attention to the two alternatives to your assigned approach.

Note that your group may have uncovered a novel solution not included in this list of three. You may choose to work through the structured decision matrix with your self-identified solution as a fourth solution option.

2.5.2 Additional Sources

1. **Planting ground cover :** A novel approach on the delineation of a multipurpose energy-greenbelt to produce biofuel and combat desertification in arid regions
2. **Supporting local farmers:** How can Africa farmers protect against desertification | World Economic Forum (weforum.org)
3. **Adopting innovative water management strategies**: Bunds are doing their job: re-greening Kenya and reducing flooding | Dutch Water Sector

Score each of your three potential solutions for each of the desired outcomes using a simple relative scale of 1 for least benefit to 5 for greatest benefit. Using a relative scale means you don't need to know exactly how well each solu-tion meets the goal of each desired outcome, but you can use your judgment to assess how well each solution works compared to the others.

While this is a relative (subjective) scale, note that you will need to justify your scoring of each solution for each desired outcome.

Once each cell in your decision matrix has a relative score, calculate an average score for each solution. Based on this analysis, which is the "best" solution, considering all your desired outcomes?

2.5.3 Role Playing

(Key Skill: Communicating Science)

In tackling environmental issues, you'll often find yourself working with groups of people with different perspectives about implementing a solution. You should use clear, concise communication to summarize your solution and justify its selection. To be effective, you must address concerns likely to be presented by various stakeholder groups and how the risks of taking these actions outweigh the risks of taking no action.

Your group will present and justify your chosen solution to the class with a particular emphasis on how it might benefit or impact one of the key stakeholder groups listed below. Be sure to include arguments that support this solution from the ecological, societal, and economic domains of sustainability science. Listeners will also be **assigned a stakeholder identity**, with an opportunity to **ask follow-up questions** after your presentation that reflect their unique concerns and perspectives. Your job will be to listen carefully and tailor your answers to this audience of stakeholders.

Key stakeholder groups include the following:

- Local residents who will adopt the new techniques
- Extension officers who educate local residents about how to implement changes
- Community farmers focused on maximizing agricultural yields
- Representatives of NGOs providing the funding for the work

2.5.4 Reflecting on Your Work

(Key Skill: Personal Reflection)

During your work in Solutions, you have explored several possible actions that could be taken to help stabilize the health of soils in sub-Saharan Africa and slow the process of desertification. Take a little more time now to reflect on your findings and the skills you practiced. Consider the following prompts but feel free to expand on any of them to best capture your learning experience and feelings about this issue.

- Reflect on your work through a sustainability science lens. Does your solution address ecological, economic, and societal considerations? Which considerations do you think should carry the most weight in the decision? Why?
- How did you weigh solutions that might have the greatest environmental impact against those that are most likely to be implemented and maintained for long-term impact?
- How can environmental scientists work to show the value of healthy ecosystems and justify the costs of mitigation strategies?
- Science communication can be challenging, especially when working with diverse audiences. We need to craft our communication to match the interests and values of the target audience, but how do you do this when your audience contains a mix of stakeholder groups? How can you maximize the impact of your message to a diverse audience?

Unit Solutions Summary *Summarize and justify your final solution choice and outline how it addresses the direct challenge while also considering social, economic, and ecological impacts. Also demonstrate that it will continue to meet the challenges posed by climate change.*

2.6   Desertification: Final Challenge

As a part of this Unit Challenge, you were asked to write a one-page Fact Sheet justifying your choice for an approach to slow the process of desertification at as many individual farms in sub-Saharan Africa as possible. Your Fact Sheet should include the following components:

- Brief problem statement
- Recommended mitigation strategy with sufficient details to summarize the general approach
- Justification of this recommendation (e.g., long-term effectiveness given anticipated climate changes, implementation costs, other benefits provided, etc.). Be sure to use a sustainability lens to include considerations of direct and indirect ecological, social, and economic considerations
- Any obstacles the group might face trying to implement your solution

Consider using figures, graphics, and tables to help summarize the system and show how this solution is well suited to meet all desired outcomes.

Final Unit Challenge Submit your final recommendations in a one-page Fact Sheet using clear science communication designed for a lay audience.

References

Agula C, Akudugu MA, Dittoh S, Mabe FN (2018) Promoting sustainable agriculture in Africa through ecosystem-based farm management practices: evidence from Ghana. Agric Food Secur 7(1):1–11

Blunden J, Arndt DS (eds) (2020). State of the climate in 2019. Bull Am Meteorol Soc 101(8):S1–S429

Integrated Panel on Climate Change (IPCC) (2014) Climate change 2014 – impacts, adaptation and vulnerability: regional aspects. IPCC, Cambridge University Press, Cambridge

Integrated Panel on Climate Change (IPCC) (2019) Summary for policymakers. In: Climate change and land: an IPCC special report on climate change, desertification, land degradation, sustainable land management, food security, and greenhouse gas fluxes in terrestrial ecosystems (Shukla PR, Skea J, Calvo Buendia E, Masson-Delmotte V, Pörtner HO, Roberts DC, Zhai P, Slade R, Connors S, van Diemen R, Ferrat M, Haughey E, Luz S, Neogi S, Pathak M, Petzold J, Portugal Pereira J, Vyas P, Huntley E, Kissick K, Belkacemi M, Malley J (eds))

Tedesse H (2014) Suitability analysis for *Jatropha curcas* production in Ethiopia-a spatial modeling approach. Environ Syst Res 3(1):1–13

World Hunger Education Service (2018) Africa hunger and poverty facts. https://www.worldhunger.org/africa-hunger-poverty-facts-2018/

Zougmore R, Abdulai J, Tioro A (2014) Climate-smart soil water and nutrient management options in semiarid West Africa: a review of evidence and analysis of stone bunds and zaï techniques. Agric Food Secur 3(1):16. https://doi.org/10.1186/2048-7010-3-16

Core Knowledge
Biodiversity, Ocean systems, Conservation biology

3.1 Environmental Issue

Many species of sea turtles, iconic members of marine ecosystems, are currently listed as endangered or threatened. These important keystone creatures face multiple threats, including coastal development, ocean plastic waste, and water pollution. Climate change also threatens sea turtles in several important ways. In addition to the rising sea level's impact on critical nesting habitats (Fig. 3.1), a warming climate may also result in severely imbalanced sex ratios in hatchlings, with potentially catastrophic implications for the long-term survival of the species.

Conservation biologists across the globe have been studying this unique impact of climate change on sea turtle populations and evaluating possible mitigation strategies to help maintain populations of these creatures. In this unit, you'll look at the relationship between climate change and these important marine icons and explore how local actions can help lower the risk.

3.2 Background Information

3.2.1 The Problem

Sea turtles (Fig. 3.2) are important members of marine ecosystems. These charismatic keystone species share the "celebrity" status of other highly visible species like polar bears and the American eagle. Historically, many indigenous cultures have attached great spiritual significance to the sea turtle. Today, sea turtles represent a substantial source of income from ecotourism. Moreover, of course, as keystone species, the turtles play important roles in the structure and function of marine ecosystems. They help maintain healthy seagrass beds and coral reefs, thus supporting important habitats for other marine life and facilitating nutrient cycling. They even help keep jellyfish populations in check.

Despite a widespread recognition of the importance of these marine creatures, the globe's seven sea turtle species are all listed on the International Union for Conservation of Nature's (IUCN) Red List of Threatened Species:

- Leatherbacks: Vulnerable
- Greens: Endangered
- Loggerheads: Vulnerable
- Hawksbills: Critically Endangered
- Olive Ridleys: Vulnerable
- Kemp's Ridleys: Critically Endangered
- Flatbacks: Data Deficient

Sea turtles are threatened by a host of human activities, including coastal development, plastics, vessel strikes, bycatch, illegal poaching, habitat destruction, invasive species, and water pollution. Sea turtles are especially susceptible to light pollution and degradation of nesting habitats, which can interfere with egg-laying. In addition to this long list of threats, sea turtles are also uniquely affected by climate change, with direct (hatchling sex ratios) and indirect (sea level rise threatening nesting sites) impacts.

3.2.2 The Role of Climate Change

Global climate change has introduced a new series of risks for the world's sea turtle populations. Among the climate-related challenges are the following:

- Rising sea water temperatures threaten the future of sea turtle habitats like coral reefs. It is also likely that warmer water temperatures will reduce food availability, resulting in less energy available for producing offspring.

J. Pontius, A. McIntosh, *Environmental Problem Solving in an Age of Climate Change*, Springer Textbooks in Earth Sciences, Geography and Environment, https://doi.org/10.1007/978-3-031-48762-0_3

Fig. 3.1 Sea turtle hatchlings head for the sea. (Source: Pixy.org (CC0) {Public Domain})

Fig. 3.2 Sea Turtle. (Source: Wexor Tmg (CC BY 2.0) via Wikimedia Commons)

- Sea level rise, already occurring, will continue to reduce the amount of space available for sea turtle nests on sandy beaches, perhaps forcing them to seek alternative sites in less desirable areas.
- More severe storms expected to occur in a warmer climate may increase the rate of beach erosion and even lead to flooding of some nest sites.
- Sea turtles rely on ocean currents for migration. As these currents are altered by changing global weather patterns, sea turtles may be exposed to new predators and may have to alter their range and the timing of egg laying.

However, warming temperatures also impact the long-term reproductive success of sea turtles in unique ways. These ancient reptiles, with lifespans of 50–100 years, differ from most other species in that gender is determined during embryo development instead of during fertilization. In a process known as temperature-dependent sex determination (TSD), the temperature of the developing eggs determines if hatchlings will be male or female.

Each species has a "pivotal" temperature where the result is an equal proportion of male and female hatchlings. The warmer the egg, the more likely a female will develop. While a higher female-skewed sex ratio is not necessarily bad, some recent studies in areas with sharply elevated temperatures indicate a nearly complete absence of male hatchings, with obvious implications for the long-term success of populations, especially with estimates that for every 1000 sea turtle eggs laid, roughly one turtle reaches adulthood.

This combination of longevity and TSD makes species like sea turtles particularly sensitive to climate change. Long-term survival of these species may depend on traits and behaviors that might evolve to counteract the impact of climate change (Blechschmidt et al. 2020). This could include an evolutionary advantage for turtles which dig deeper nests or favor populations that return to cooler, more shaded nesting sites (Fig. 3.3).

3.2.3   Solutions

As with most of the problems you'll consider in the units in this book, many of the threats facing sea turtles are very challenging to manage, with multiple actions required to successfully maintain viable populations. For example, impacts of coastal development can be minimized with strategic habitat protection efforts and nesting exclusion zones, while new technologies can help limit nest predation. Many governments are enacting various plastic reduction policies to minimize the impacts of plastic pollution.

However, climate change will have many direct and indirect impacts on sea turtles themselves, the habitats they depend on, and the ecosystems they are a part of. As noted earlier, elevated temperatures threaten the balance of the sexes in sea turtles. Conservation biologists have been testing several possible solutions to the threat posed to long-term reproductive success by the severely imbalanced sex ratios resulting from warming temperatures.

Most nesting beaches support vegetation that can provide various degrees of shade. The mean temperature of nests with even partial shade is typically 1 °C cooler than nests in the sun (Wood et al. 2014). Some scientists are currently testing the effectiveness of installing artificial shade structures and even nest watering to help lower nest temperature (Hill et al. 2015). Other conservation efforts include collecting sea turtle eggs (Fig. 3.4) and transporting them to artificial incubation sites where the temperature can be carefully regulated. Similarly, as has been done during other environmental crises like the Wide Horizon oil spill in the Gulf of Mexico, nests may be relocated to improve hatching success.

Fig. 3.3 A female green sea turtle nests under cover of vegetation on Heron Island in the Coral Sea. (Source: GretelW (CC BY-SA 4.0) via Wikimedia Commons)

Fig. 3.4 Conservationists work to evacuate a sea turtle nest. (Source: Hillebrand S., US FWS {Public Domain})

3.2.4 Unit Challenge

The overarching challenge you face is to determine how to ensure the ongoing development of a reproductively successful ratio of male to female sea turtles. Specifically, this challenge focuses on leatherback sea turtle populations at a site in Costa Rica (Fig. 3.5). At this location, low levels of precipitation and high temperatures have been linked to decreased hatching success and emergence of hatchlings from the nest (Hill et al. 2015). Your Unit Challenge is **to identify the best way to reduce nesting temperatures and ensure that there are an adequate number of male individuals to enable long-term breeding success and ultimate population survival**.

The exercises in this unit will help you develop a management plan for a turtle conservation group in Costa Rica whose goal is to bring sex ratios among yearly turtle hatchlings into balance by achieving a 1 °C reduction in nesting temperatures at Playa Grande, Costa Rica.

The Scenario

Working to successfully manage sea turtle populations requires both a knowledge of the basic biology of the creature and an understanding of how they're responding to climate change. After familiarizing yourself with the Playa Grand ecosystem and consulting the literature on sea turtle nesting site mitigation measures, you've chosen to analyze and compare these three options:

1. **Lower air temperatures above sea turtle nests.** One obvious way to lower the temperatures of sea turtle nests is to cool the overlying air to the extent possible. Working on the beaches of Playa Grande, Hill et al. (2015) demonstrated that shading could substantially reduce sand temperatures.

   Specific recommendation: Install shade tents over sea turtle nesting sites.

2. **Cool nesting sites using water.** Turtle experts believe that irrigating nests can cool sand temperatures and increase the number of male offspring. Sources of freshwater, however, may be limited, particularly on beaches in remote areas. As an alternative, Smith et al. (2021) treated nests with seawater and found that even one application could help create male turtle hatchlings.

   Specific recommendation: Install irrigation drip lines to turtle nesting sites.

3. **Relocate nests to greater depths**. As one digs more deeply into sand, temperatures cool. Increasing the depth of sea turtle nests may offset the effects of climate-related temperature increases. One concern is that the increased precipitation expected with climate change may raise water table levels and increase egg mortality in deeper nests (Rivas et al. 2018).

   Specific recommendation: Relocate existing sea turtle nests to greater depths.

Fig. 3.5 At Playa Grande National Park, leatherback sea turtles that hatched at Playa Grande have returned for thousands of years to lay their eggs. (Source: Jorge Antonio Leoni de León (CC BY SA 4.0) via Wikimedia Commons)

Each of these options has various pros and cons and different costs and benefits. Your job is to evaluate these options and recommend a practice that will perform best over time under the conditions expected with climate change.

3.2.5   Relevant Facts and Assumptions

- Each nest contains approximately 100 eggs. You expect to be able to identify approximately 1000 nests over the breeding season.
- Your goal is to treat as many nests as possible with your chosen mitigation efforts.
- For each sea turtle nest, you can install and maintain shade tents for $20 US; you can install 10 tents per day.
- Installing and maintaining irrigation drip lines costs $25 US per nest; systems can be installed at 7 nests daily.
- Relocating nests to deeper sites costs $15 US per nest; 15 nests can be relocated each day.
- You have an annual budget of $20,000 US from a Turtle Island Restoration Network grant to support your efforts.
- Local labor is available to perform the necessary work.

3.2.6   Build Your Foundational Knowledge

Below are web sources that provide additional information about each of the solutions you're considering for this Unit Challenge. This information can build a critical foundation to help you evaluate each option and support your final choice. After reviewing each source, be prepared to answer questions in the Preparation Assessment Quiz and to summarize any information relevant to your Unit Challenge.

Shade Structures (Costa Rica):
Totally cool turtles may help save species

Shade and Irrigation (Columbia):
Controlling Sand Temperatures for Sea Turtles in the Choco Region of Colombia

Relocation and Irrigation (Florida):
It's Gettin' Hot in Here, So Water All Your Eggs

Final Product: A one-page Fact Sheet summarizing the issue, detailing your solution, and justifying your choice of that solution. Consider your audience, land managers, community members, and conservation organizations at Playa Grande, Costa Rica. Be sure to demonstrate how your proposed solution will stand up to the ongoing challenges posed by climate change.

3.2.7   Preparation Assessment Quiz

Are you ready to tackle your challenge? At this point you should understand the basic environmental principles and ecological processes involved in this environmental problem. Consider the following questions. If you are comfortable with answering these, then you are ready to head into Discovery, Analysis, and Solutions activities.

- Why are sea turtles considered keystone species?
- What are some of the ways that human activities harm sea turtles?
- What is temperature-dependent sex determination?
- Why is a female-skewed sex ratio not necessarily a bad thing for most species?
- The University of Queensland study suggested that if resources to irrigate the nests were lacking, a shade structure could be made of which two materials?
- According to the World Wildlife Federation, the research on sea turtles in the Choco Region of Columbia involved how many measurements of sand temperature?

- Brandon Wei's article notes that sea turtle eggs are not like chicken eggs. What are the two differences?
- For each of your proposed solutions, are there any additional benefits that might arise from their implementation that might not be directly related to your Unit Challenge?
- For each of your proposed solutions, are there any negative unintended consequences that might result from their implementation?
- What additional information did you glean from your web sources that might help inform your Unit Challenge?

3.3 Sea Turtles: Discovery

Specific Skills You'll Need to Review: Navigating the Scientific Literature, Science Communication, Problem-Solving

3.3.1 Independent Research

(Key Skill: Navigating the Scientific Literature)

To better understand the role of climate in sea turtle hatchling sex ratios and the potential impact of various mitigation activities, you first need to examine the literature to see what others have found. Conduct a search of the peer-reviewed scientific literature focused on the solution you have been assigned, and identify one research paper that focuses on your assigned approach.

Prepare a summary of the article you selected that includes the following:

- **Citation**
- **Main topic**: Stick to a few words, likely pulled from the title.
- **General summary**: A few bulleted sentences summarizing the research question it addresses and approaches it takes.
- **Methods:** How did they approach their research question?
- **Location**: Where was the work done?
- **Conclusions**: Concise list of the findings, specifically capturing the take-home message.
- **Relevance**: How might this study help inform your choice of a solution? Feel free to make a bulleted list of information you may want to include later.

3.3.2 Literature Share: Reciprocal Instruction

(Key Skill: Scientific Communication)

Share In small groups, share and critique the research article you found. Keep in mind that your peers have not read this article, and it is your job to convey the key information to them. Note the items that will be important to consider when you are designing your solution to the challenge.

Critique Evaluate how these studies might help inform your Unit Challenge. Consider the following:

- Source (Quality of the work or bias of the authors)
- Methods (Did their methods sufficiently address the research question?)
- Conclusions (Did the results justify the conclusions made?)
- Relevance (Can these findings be applied to your challenge?)

Based on the group's critiques, choose one article to share with the larger class, along with the key information that may be useful in deciding on a solution to propose.

3.3.3 Think-Compare-Share

(Key Skill: Problem-Solving)

Now that you have more information about possible solutions for this unit's challenge, you need to **develop a more formal problem definition** to guide your work throughout the rest of the exercises.

Think Start by working independently to develop a specific Problem Statement to guide the remainder of your work. Problem Statements provide the relevant information and boundaries to make the issue something you can effectively assess and tackle. The basics of a formal Problem Statement include the following:

Problem Statement: A short, concise statement summarizing the issue that includes the following:

- A **description** of the undesired condition or change that you hope to achieve (What is the actual problem?)
- **Justification** for addressing the problem (Why does this problem matter?)
- Potential **sources** or **causes** of this problem (What is the cause you need to address?)
- The **metrics** you will use to assess the status of the problem (How will you know if you are making a difference in the problem?)
- The **desired outcome** for these metrics (What is the end goal or ideal state?)

- Potential **solutions** to consider (How might you attempt to achieve this goal?)

Compare/Share Now return to your small group to share your Problem Statements. Use each of your ideas to develop a joint statement that contains all key information and is concise, clear, and well written.

Unit Discovery Summary *Submit a final Problem Statement that succinctly captures the key information to guide your work on this Unit Challenge.*

3.3.4   Reflecting on Your Work

(Key Skill: Personal Reflection)

After your work in Discovery, you should have a better idea of the problems you face and have produced a Problem Statement you can use to tackle the Unit Challenge. Take a moment to reflect on this work. Consider the following prompts but feel free to expand on any to best capture your learning experience and better inform your next steps.

- Of the skills you practiced in Discovery, which were the most challenging? Which were the most interesting?
- How were you most comfortable working during these exercises? In small groups, independently, or with the larger class? Why? How does your choice reflect your personality type and leadership style?
- Was your Problem Statement strictly focused on sea turtle conservation, or did it also consider important social and economic considerations? How might a focus on the ecological aspects limit your ability to identify truly sustainable solutions?
- You've been given three viable solutions to assess as a part of this case study. But this is not an exhaustive list of options or even necessarily the best possible course of action for every scenario. Take a moment to "think outside the box." Are there any other possible solutions you think would be worth exploring? Describe one that you think would be worth pursuing.

3.4   Sea Turtles: Analysis

Specific Skills You'll Need to Review: Quantitative Literacy, Sustainability Science

Review your Background and Discovery sections before beginning the Rotating Station exercises below. While you focused on one potential solution in your Independent Research in Discovery, keep an open mind as your work through Analysis activities.

3.4.1   Rotating Stations

(Key Skill: Quantitative Literacy)

At each of the following stations, you will review data that are relevant to the three potential solutions you're considering. Spend some time working through the analyses at each station to learn more about this issue and possible solutions for your Unit Challenge.

Be sure to write down one finding at each station that will help inform your selection of a solution.

Station 1: Shading Sea Turtle Nests Figure 3.6 (Hill et al. 2015) shows sand temperatures at sea turtle nesting sites at two different depths with shading or without shading and at three different watering levels (high, average, and low). Note how temperature varies between shaded (C, D) and exposed (A, B) nests, between shallow (A, C) and deeper (B, D) nests, and among various watering strategies (line pattern). Because all figures use the same scale and range for the x and y axes, you can draw some general conclusions about these treatments.

Examining this figure, answer the following:

- Isolate the impact of shade by comparing temperatures considering only the solid lines in panels C and D. The gray solid line had no watering but was shaded and the black solid line had no watering but was exposed. How much of an impact did shading have on temperature at the two depths?
- Now isolate the impact of depth by comparing the solid black line only for the exposed (panels A and B) and shaded (panels C and D) sites. How much of an impact did nesting depth have on temperature?
- How did the intensity of watering (high, average, and low dotted lines) impact temperature compared to nests with no watering (solid lines)?

Record any relevant Station 1 findings

Station 2: Irrigating Sea Turtle Nests Using this same figure from Hill et al. (2015), now answer the following about the interactions among these three treatments (depth, exposure and watering).

- How did watering nests differ at shallow and deeper nesting depths? At shaded and exposed sites?

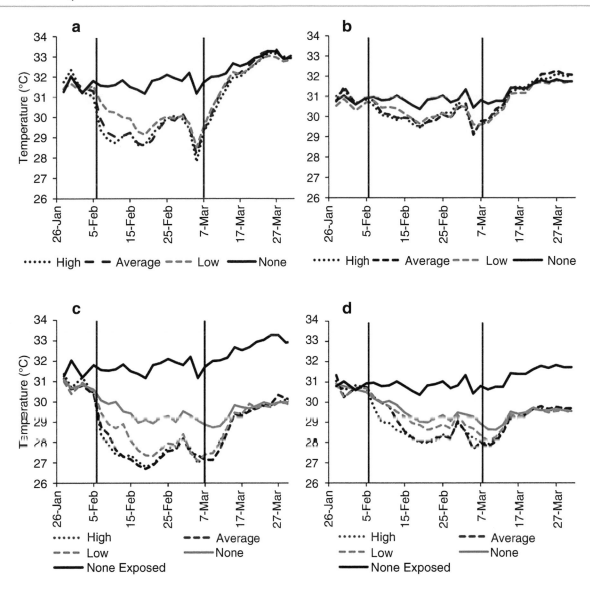

Fig. 3.6 Sand temperature (°C) through time at various turtle nest sites. (**a**) 45 cm exposed; (**b**) 75 cm exposed; (**c**) 45 cm shaded; and (**d**) 75 cm shaded, at high, average, and low water treatments. Vertical lines mark the beginning and end of watering. (Source: Hill et al. 2015 {Open Access})

- How did temperatures in the shallow and deeper nests respond once watering ended?
- Which combination of treatments (depth, shade and watering) resulted in the lowest nest temperature?

Record any relevant Station 2 findings

Station 3: Relocating Sea Turtle Nests Liles et al. (2019) measured the sex ratio and physical condition of hatchlings from nests at two locations (El Salvador and Nicaragua) under three different treatments: (a) protected *in situ*; (b) relocated to protected locations on the beach (beach); or (c) moved to vegetated hatcheries (hatchery). Figure 3.7 shows how the proportion of females, which increases with warmer nesting conditions, varied between

the two locations over time and across the three treatments.

Consider that the goal is to have close to 60% females in each clutch, thus maximizing the long-term reproductive success of the species. Based on the data presented above, answer the following questions:

- In panels (a) and (b), gray bars represent the proportion of total nests laid during each 2-week period over the season. When are most nests laid at these locations?
- How does the proportion of female hatchings vary over the season? What does this difference tell you about how the timing of nesting might impact the success of hawksbill populations?

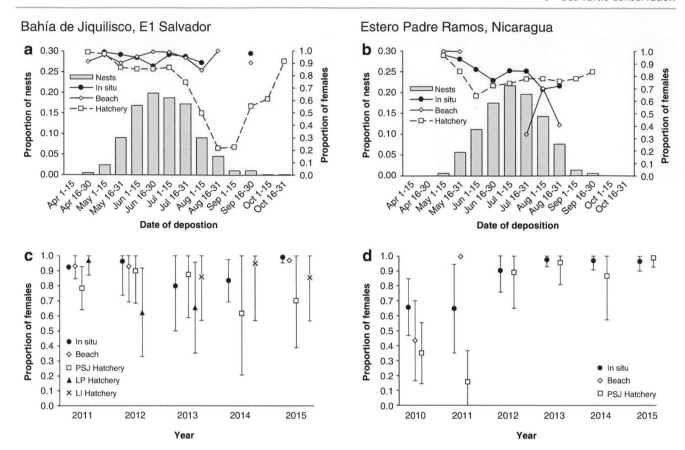

Fig. 3.7 Estimated hawksbill hatchling sex ratios at Bahía de Jiquilisco, El Salvador (2011–2015) and Estero Padre Ramos, Nicaragua (2010–2015). (**a**, **b**) Bimonthly frequency distribution of hawksbill nesting (gray bars) and estimated offspring sex ratios from three nest protection strategies (lines) at Bahía de Jiquilisco (n = 835 clutches) and Estero Padre Ramos (n = 1196 clutches), respectively. (**c**, **d**) Annual mean (±SD) estimated offspring sex ratios from each nest protection strategy at Bahía de Jiquilisco and Estero Padre Ramos, respectively. (Source: Liles et al. 2019 {Open Access})

- Compare the three treatments across all years in panels (c) and (d). Which treatment minimized the dominance of females? How consistent was this response over years?
- If the goal is to achieve a 60% female ratio in offspring, what would you suggest at each location?

Record any relevant Station 3 findings

Station 4 In your Unit Challenge, you were given a set of assumptions that included estimated costs (money and time) required to implement each of your three possible solutions. Assuming that the goal is to implement your chosen solution across as many nests as possible, answer the following:

- Assume that sea turtle eggs take about 2 months to hatch. How many nests can you treat over that 60-day period based on the time constraints stated for each proposed solution?
- Assume that you have only $20,000 budgeted for this work. How many nests could be treated using each proposed solution within that budget?

- How much should these costs weigh in the selection of any one potential solution? How do you weigh costs and number of nests you can treat against the actual effectiveness of the treatment?

Record any relevant Station 4 findings

3.4.2 System Mapping

(Key Skill: Sustainability Science)

Your assigned solution would require some physical structure or nest manipulation that will have impacts beyond the hatchlings themselves. There may be a variety of direct and indirect economic, social, and ecological impacts that should also be considered. For example, irrigation systems may impact water supply for local residents (social impact); shade tents may decrease the aesthetic appeal and recreational opportunities on the beach (economic impact); and burying nests deeper may impact native predators that rely on sea turtle eggs as a food source (ecological impact).

Fig. 3.8 Template
sustainability map to help
connect ecological, societal,
and economic considerations
associated with your potential
solutions

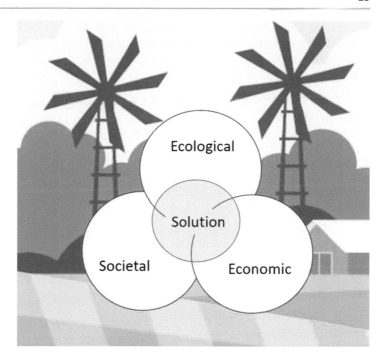

Working with other class members assigned the same solution, develop a simple sustainability map that shows the various system connections across each of the three sustainability domains (ecological, economic, and societal). When you find connections between impacts that cross domains, draw a line to identify the connection (e.g., decreased recreational opportunities (economic) leading to decreased financial resources for a family (social)).

The goal is to think broadly about this larger system and envision how implementing actions in one domain (e.g., ecological) may impact components in another domain (e.g., economic or societal).

Use the following template (Fig. 3.8) to get started:

When each group has completed their basic sustainability map, come together as a class to compare lists for the three possible solutions. This information will help inform your decision support work later in the unit.

Unit Analysis Summary   Based on your explorations, what have you learned that can help inform your choice of a solution? Do the data support the adoption of one of these potential solutions?

3.4.3   Reflecting on Your Work

(Key Skill: Personal Reflection)

During your work in Analysis, you explored some of the research into possible solutions to help inform your decision. Take a moment to reflect on this work. Consider the following prompts but feel free to expand on any to best capture your learning experience and better inform your next steps.

- How did you feel working with data? Do you consider quantitative literacy a strength or an area for improvement for you?
- How important should science be in informing management and policy? Do you feel the data you examined support and justify the costs of addressing your challenge?
- Any solution should be examined using a sustainability science lens. Your goal is to work with conservationists to protect sea turtle populations in Costa Rica. But any solution you implement may have other direct and indirect impacts. What are other possible economic, social, or ecological impacts? How should these considerations influence your decision?

3.5   Sea Turtles: Solutions

Specific Skills You'll Need to Review: Problem-Solving, Decision Support, Communicating Science

Review your Unit Challenge and the major findings from Discovery and Analysis, including the sustainability map you created to highlight connections among possible solutions and the larger socioecological system

3.5.1   Small Group Guided Worksheets

(Key Skill: Decision Support)

Decision support matrices can help break down the desired outcomes to reflect multiple criteria for consideration, and they allow you to compare how each possible solution achieves those desired outcomes. This not only helps inform decision-making, but it also provides transparency in the decision process and justification to help you advocate for adopting a particular solution.

For the three possible solutions, you will evaluate how well each can achieve the following desired outcomes:

- Effectively reduces nest temperatures
- Maximizes the number of nests treated with available resources
- Minimizes costs of implementation
- Provides a lasting impact (not a short-term fix), even with a changing climate
- Is widely accepted by stakeholder groups
- Minimizes secondary impacts (e.g., to other animal species, local economies)

Considering your three potential solutions, develop a formal decision support matrix to compare and evaluate each approach using the template matrix below (Table 3.1). Getting the most out of the decision matrix requires a depth of knowledge about each of the three possible solutions. Below we list additional sources about each option. Please review each of these, paying particular attention to the two alternatives to your assigned approach.

Note that your group may have uncovered a novel solution not included in this list of three. You may choose to work through the structured decision matrix with your self-identified solution as a fourth solution option.

3.5.2 Additional Sources

1. **Shading sea turtle nests**: 'Nest Shading' Conservation Technique Must Be Carefully Deployed in Order to Protect Sea Turtle Populations, According to New Study – Drexel News Blog
2. **Irrigating nests**: Researchers Are Cooling Down Turtle Nests for Conservation Purposes | Plants And Animals (labroots.com)
3. **Relocating nests:** Relocating a Sea Turtle Nest – Ocean Pancake

Score each of your three solutions for each of the desired outcomes using a simple relative scale of 1 for least benefit to 5 for greatest benefit. Using a relative scale means you don't need to know exactly how well each solution meets the goal of each desired outcome, but you can use your judgment to assess how well each solution works compared to the others.

While this is a relative (subjective) scale, note that you will need to justify your scoring of each solution for each desired outcome.

Once each cell in your decision matrix has a relative score, calculate an average score for each solution. Based on

Table 3.1 Basic decision matrix to compare the three possible solutions for your Unit Challenge. Include your justification for each solution's ranking for each desired outcome

| Solution Options | **Desired Outcomes** | | | | | |
	Effectiveness (temperature reduction)	**Maximizes the #nests treated**	**Minimizes installation cost**	**Maximizes duration / longevity of effect**	**Maximizes stakeholder buy-in**	**Minimizes secondary impacts**
Install shade structure						
Install irrigation systems						
Move nests to deeper holes						
Justification for Ranking						

Use a scale of 1-5 to score each option under each desired outcome category where 1 = does not achieve the desired outcome and 5 = completely achieves the desired outcome.

this analysis, which is the "best" solution, considering all your desired outcomes?

3.5.3 Role Playing

(Key Skill: Communicating Science)

In tackling environmental issues, you'll often find yourself working with groups of people with different perspectives about implementing a solution. You should use clear, concise communication to summarize your solution and justify its selection. To be effective, you must address concerns likely to be presented by various stakeholder groups and how the risks of taking these actions outweigh the risks of taking no action.

Your group will present and justify your chosen solution to the class with a particular emphasis on how it might benefit or impact one of the key stakeholder groups listed below. Be sure to include arguments that support this solution from the ecological, societal, and economic domains of sustainability science. Listeners will also be **assigned a stakeholder identity**, with an opportunity to **ask follow-up questions** after your presentation that reflect their unique concerns and perspectives. Your job will be to listen carefully and tailor your answers to this audience of stakeholders.

Key stakeholder groups include the following:

- Local landowners who want to maintain the pristine aesthetic of the beach
- Politicians interested in local economies and businesses
- Conservationists worried about long-term survival of local sea turtle populations
- Scientists who advise that any interventions may interfere with natural adaptation to new climate conditions

3.5.4 Reflecting on Your Work

(Key Skill: Personal Reflection)

During your work in Solutions, you have explored several possible actions that could be taken to cool nesting sites and prevent an imbalance in sea turtle hatchling sex ratios. Take a little more time now to reflect on your findings and the skills you practiced. Consider the following prompts but feel free to expand on any of them to best capture your learning experience and feelings about this issue.

- Reflect on your work today through a sustainability science lens. Does your solution address ecological, economic, and societal considerations? Which considerations

do you think should carry the most weight in the decision? Why?
- How did you weigh solutions that might have the greatest environmental impact against those that are most likely to be implemented and maintained for long-term impact?
- How can environmental scientists work to show the value of healthy ecosystems and justify the costs of mitigation strategies?
- Science communication can be challenging, especially when working with diverse audiences. We need to craft our communication to match the interests and values of the target audience, but how do you do this when your audience contains a mix of stakeholder groups? How can you maximize the impact of your message to a diverse audience?

Unit Solutions Summary *Summarize and justify your final solution choice and outline how it addresses the direct challenge while also considering social, economic, and ecological impacts. Also demonstrate that it will continue to meet the challenges posed by climate change.*

3.6 Sea Turtles: Final Challenge

As a part of this Unit Challenge, you were asked to write a one-page Fact Sheet recommending a management plan to bring sex ratios in sea turtle hatchlings closer to historical norms by reducing nest temperatures for as many sites as possible. Your Fact Sheet should include the following components:

- Brief problem statement
- Recommended mitigation strategy with sufficient details to summarize the general approach
- Justification of this recommendation (e.g., long-term effectiveness given anticipated climate changes, implementation costs, other benefits provided, etc.). Be sure to use a sustainability lens to include considerations of direct and indirect ecological, social, and economic considerations
- Any obstacles the group might face trying to implement your solution

Consider using figures, graphics, and tables to help summarize the system and show how this solution is well suited to meet all desired outcomes.

Final Unit Challenge Submit your final recommendations in a one-page Fact Sheet using clear science communication designed for a lay audience.

References

Blechschmidt J, Wittmann MJ, Blüml C (2020) Climate change and green sea turtle sex ratio – preventing possible extinction. Genes 11(5):588

Hill JE, Paladino FV, Spotila JR, Tomillo PS (2015) Shading and watering as a tool to mitigate the impacts of climate change in sea turtle nests. PLoS One 10(6):e0129528

Liles MJ, Peterson TR, Seminoff JA, Gaos AR, Altamirano E, Henríquez AV, Gadea V, Chavarría S, Urteaga J, Wallace BP, Peterson MJ (2019) Potential limitations of behavioral plasticity and the role of egg relocation in climate change mitigation for a thermally sensitive endangered species. Ecol Evol 9(4):1603–1622

Rivas ML, Spinola M, Arrieta H, Faife-Cabrera M (2018) Effect of extreme climate events resulting in prolonged precipitation on the reproductive output of sea turtles. Anim Conserv 21(5):387–395

Smith CE, Booth DT, Crosby A, Miller JD, Staines MN, Versace H, Madden-Hot CA (2021) Trialing seawater irrigation to combat the high nest temperature feminization of green turtle *Chelonia mydas* hatchlings. Mar Ecol Prog Ser 667:177–190. https://doi.org/10.3354/meps13721

Wood A, Booth DT, Limpus CJ (2014) Sun exposure, nest temperature and loggerhead turtle hatchlings: implications for beach shading management strategies at sea turtle rookeries. J Exp Mar Biol Ecol 451:105–114

Core Knowledge

Water quality, Marine and freshwater ecosystems, Conservation, Biology, Stormwater runoff

4.1 Environmental Issue

Puget Sound, a complex 2600 km² inlet of the Pacific Ocean and part of the Salish Sea, is located along the northwestern coast of the State of Washington. Economically important as a port, a shipping route, and an active commercial fishery and host for a broad array of tourism and recreational activities, Puget Sound also supports a vibrant and diverse marine ecosystem. This includes commercially important fish species like herring and smelt, the geoduck, a clam prized in Asia, and a number of threatened and endangered species, including the iconic orca or killer whale (*Orcinus orca*) (Fig. 4.1) and the Chinook salmon (*Oncorhynchus tshawytscha*).

However, human activities and continued population growth within the basin (Fig. 4.2) have degraded water quality, marine habitats, and the natural resources which support the ecosystem. The sustainability of this unique system, with its mix of terrestrial, freshwater, estuarine, and marine components, is currently threatened by a host of environmental stressors, with climate change further increasing pressure on the Sound and the communities it supports. In this unit, you'll first explore the stressors that threaten the Sound and then consider some specific steps to address one important concern: stormwater runoff.

4.2 Background Information

4.2.1 The Problem

Signs of decline in the Puget Sound socioecological system have been apparent for decades. A 1986 report from the Puget Sound Water Quality Authority stated that the Sound "…has been transformed from a relatively pristine body of water to the center of a bustling, heavily populated area. Human activity has damaged the Sound—from destroying over half its wetlands to contaminating its bottom sediments" (The Puget Sound Water Quality Authority 1986).

Many of the challenges facing the Puget Sound and the broader Salish Sea can be traced back to the more than 4 million residents living in its drainage basin and its heavy use for commercial and recreational purposes (Problems of the Puget Sound). Increased population growth in the region has contributed to an influx of chemical pollutants, including both toxic chemicals and nutrients, silt and sediment, and invasive species (Fig. 4.3).

The Sound is home to a diverse mosaic of habitats and species (Habitats of the Puget Sound watershed). Of particular interest is the status of key Puget Sound species like the Chinook salmon in the Duwamish/Green River. The Duwamish is the name given to the lower 19 km of the Green River (Fig. 4.4). It flows over 150 km from the Cascade Mountains to Elliott Bay in Puget Sound, providing critical habitat for a variety of native fish species (Fisheries of the Green-Duwamish River watershed). The largest of the Pacific salmon, Chinook migrate from Puget Sound to spawn on gravel streambeds in the Green River. While much of their historical spawning grounds are in the upper Green River sub-watershed, the Howard Hanson dam now restricts salmon spawning to the lower sub-watersheds of the Green River.

Sediments entering the river basin threaten salmon populations. Research has shown that total suspended sediments (TSS) can impact salmon in several ways. Suspended sediments not only cause physiological effects like gill trauma and changes in blood chemistry but can also lead to avoidance behaviors and territoriality and act as impediments to homing and migration (Bash et al. 2001). Many of these impacts are synergistic, with one effect leading to a host of other impacts that may affect the growth, reproduction, and survival of the fish.

© The Author(s), under exclusive license to Springer Nature Switzerland AG 2024
J. Pontius, A. McIntosh, *Environmental Problem Solving in an Age of Climate Change*, Springer Textbooks in Earth Sciences, Geography and Environment, https://doi.org/10.1007/978-3-031-48762-0_4

Fig. 4.1 *Orca* in Puget Sound. (Source: Mike Charest (CC BY 2.0) via Flickr)

An important source of TSS in the Duwamish/Green River system is stormwater runoff. Every time it rains 2.5 cm in Seattle, about 100,000 L of storm water are generated by a 0.4 hectare (ha) paved parking lot. A 110 m² roof generates 2800 L of storm water from that same amount of rainfall. While TSS levels average about 150 mg/L in urban runoff, sites with exposed soils, like construction zones, can exceed levels of 1000 mg/L.

4.2.2   The Role of Climate Change

Over the past several decades, climate change has worsened many of the problems faced by Puget Sound and added new pressures to this critical socioecological system. A report by the Climate Impacts Group at the University of Washington (Mauger et al. 2015) identified a number of direct challenges facing the Sound as the climate changes, including the following projections of conditions in 2100: a 1.6–3 °C rise in air temperatures; a 1–1.37 m rise in sea level; and highly variable precipitation patterns leading to increased landslides and transport of sediments to surface waters (Fig. 4.5). These climate-induced changes could impact the system in a variety of ways, from increased wildfires and greater storm water inputs to surface waters to increased urban flooding and pest/disease outbreaks.

With these and other direct and indirect effects to consider, scientists and others working to protect and improve conditions in Puget Sound will be faced with a variety of challenges that will need to be tackled on a system-wide basis. To be successful, any solutions will need to consider how the changing climate may worsen the effects of existing stress agents.

4.2.3   Solutions

Fortunately, Puget Sound has benefited from the efforts of large interdisciplinary teams of scientists, engineers, managers, and the general public to protect this vital resource. These groups will need to redouble their efforts as the climate changes. There are several different issues these groups are currently focused on:

Controlling Pollutant Sources With the Sound being viewed as a valued resource throughout the basin, its overall condition has been regularly evaluated by local, state, and federal managers and scientists. Key hot spots containing high levels of toxic contaminants found in the sediments of the lower reaches of Seattle's Duwamish River have been designated as US EPA Superfund sites targeted for cleanup. Cities like Seattle have been at the forefront of efforts to control pollutant sources from stormwater runoff. However, the basic problem regarding source control is that the Sound receives much of what the basin generates, including a host of pollutants from non-point sources throughout the watershed.

Protecting Species at Risk According to Zier and Gaydos (2016), as of 2015, there were 125 species of marine life considered to be at risk in the Salish Sea, with the number increasing at a rate of 2.6% annually. As a result, there are a number of programs underway to both reduce inputs of pollutants that directly threaten marine life and restore the habitats of the species at risk (More Info on Puget Sound Implementation Strategies). Ongoing efforts to help restore and protect specific populations of important marine species in Puget Sound include the following:

Fig. 4.2 The Puget Sound watershed. (Source: USGS via Wikimedia Commons {Public Domain})

Chinook salmon: Numbers of Chinook salmon have plummeted in Puget Sound in recent years, and Chinook are now considered an endangered species. Chinook salmon are not only prized by commercial and sport anglers, but they are also a key food source for orcas. Restoration strategies include improving degraded habitats of salmon and their prey species such as smelt, reducing pollutant-containing stormwater discharges, and dredg-

Fig. 4.3 The Shell Puget Sound oil refinery. (Source: Walter Siegmund (CC BY SA 3.0) via Wikimedia Commons)

ing and removing historically contaminated sediments. Other efforts to raise and release Chinook salmon are spearheaded by the US Fish and Wildlife Service (Fig. 4.6).

Orcas/Killer whales: The Southern Resident orcas reside in Puget Sound and the waters of the Salish Sea from late spring through the fall. The numbers of these migrants have been steadily declining, resulting in the listing of orcas as endangered in 2005. While a number of factors have contributed to the demise of the orcas, the most often cited are a lack of prey (they feed almost exclusively on Chinook salmon), vessel noise pollution and traffic, and degraded water quality (e.g., toxic contaminants) (the Orca Task Force not to the rescue).

Efforts to protect and restore key Puget Sound species at risk are extensive and admirable. However, given the complex challenges that climate change will pose to the ecosystem, a more holistic approach will likely be necessary to prevent further declines in the health of the Puget Sound ecosystem. You'll consider some possible approaches as you work through the exercises in this unit.

4.2.4   Unit Challenge

The overarching challenge you face is to determine how to best protect Puget Sound's iconic marine species, currently under threat from various stressors, ranging from noise pollution to stormwater runoff to contaminated sediments. Many of these threats will likely become more severe as climate change continues. However, successfully tackling such complex issues often requires that you focus on more tangible, targeted outcomes. Specifically, **your Unit Challenge is to reduce the amount of total suspended solids (TSS) entering the Green-Duwamish River**, a critical spawning ground for Chinook salmon.

One of the climate-related changes likely to impact the quality of the rivers and streams draining into Puget Sound and the important species like the Chinook salmon that rely on them will be more intense storm events and the high levels of TSS in runoff that result. After entering a stream or river, these particles can settle onto and cover the gravel on river and stream bottoms where salmon lay their eggs. The sediment particles can also clog and damage the gills of salmon and transport a number of potentially toxic pollutants into the river.

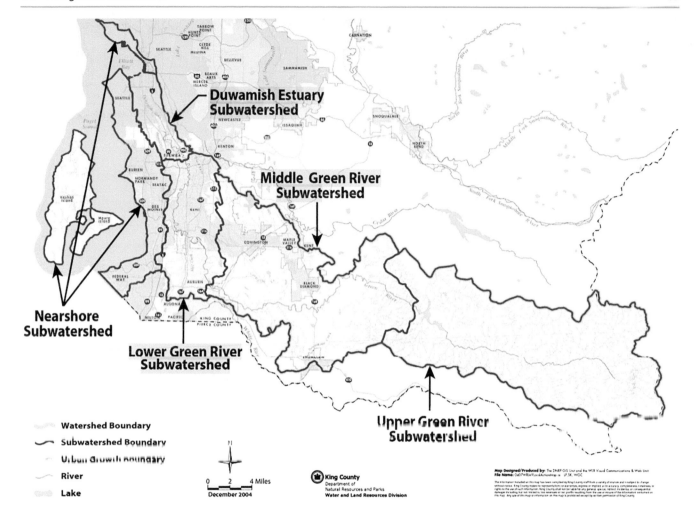

Fig. 4.4 The Duwamish watershed is located in the highly industrialized areas of Seattle, WA. (Source: Govlink.org {Public Domain})

4.2.5 The Scenario

You are faced with a difficult challenge: protecting an already-endangered iconic fish species from the risks posed by a climate future characterized by more intense rain events. When considering your options, you will evaluate a number of possible approaches. You want to make a choice that will do the most to reduce TSS carried by stormwater discharges into the Duwamish River while avoiding secondary impacts from your chosen solution.

You identify three potential approaches to evaluate:

1. **Stormwater detention basins**. An effective way to remove some of the pollutant load carried in stormwater runoff is to slow the velocity of the flowing runoff. This will result in the settling and removal of both the sediments in the runoff and any adsorbed pollutants. While not necessarily effective at removing dissolved pollutants, detention basins can effectively reduce levels of solid-associated substances like phosphorus.

Specific recommendation: Construct a series of stormwater detention basins along the Lower Duwamish River in Seattle to trap sediments carried by storm water.

2. **Rain gardens.** Another approach for slowing stormwater flow and promoting pollutant removal is to plant small areas in residential neighborhoods where slope creates a depression in the landscape. Planting shrubs, perennials, and flowers not only contributes to the aesthetic quality of the landscape but also serves to promote pollutant removal as vegetation will not only slow the flow and help remove TSS but also take up nutrients like phosphorus.

Specific recommendation: Plant urban rain gardens throughout Seattle to reduce movement of TSS into rivers during storm events.

3. **Porous pavement.** Traditional pavements, like most impervious surfaces, do little to slow the movement of storm water and the pollutants it carries. In fact, because particles accumulate on pavements between rain events, they can add to the pollutant burden carried in the runoff. Alternative paving materials, such as pervious asphalt and plastic grid pavers,

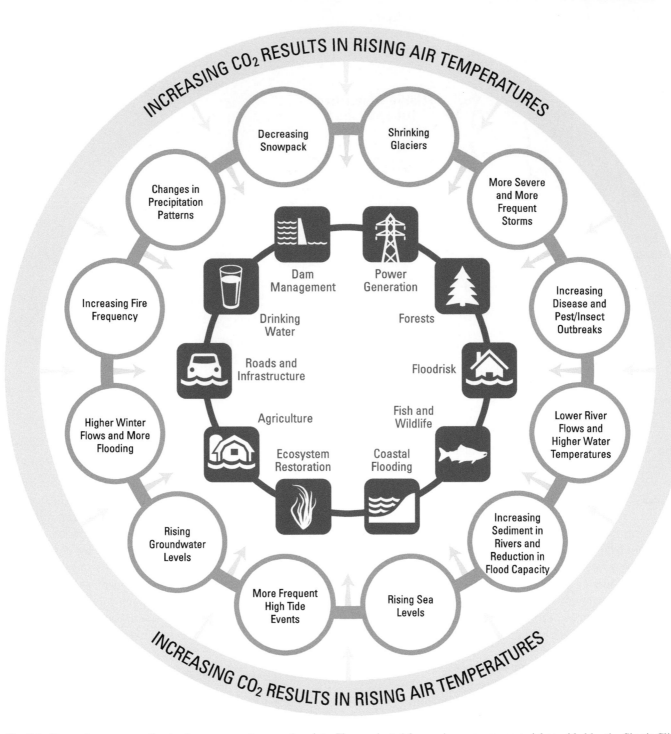

Fig. 4.5 Connections among climate change, ecosystems, and society. Figure adapted from primary source material provided by the Skagit Climate Science Consortium (SC²). For more information, visit www.skagitclimatescience.org. (Source: Mauger et al. 2015 (CC BY SA 3.0))

Fig. 4.6 Winthrop National Fish Hatchery spring Chinook and Coho smolts. (Source: USFWS/Courtesy of Yakama Nation Fisheries via Flickr {Public Domain})

allow precipitation to seep through the surface to underlying layers, filtering out pollutants in the process.

Specific recommendation: Replace Seattle's paved public parking lots with porous pavement.

Each of these three approaches has important advantages and disadvantages and costs and benefits. Your task is to evaluate each and recommend the approach that will, in your opinion, perform best at reducing TSS and supporting Chinook salmon as the climate continues to change. As you consider the three options presented above, make sure you consider the ability of the approach to reduce TSS inputs, its long-term performance under changing climatic conditions, its public acceptance, and its costs vs benefits.

4.2.6 Relevant Facts and Assumptions

- There are four locations along the Lower Duwamish River in the greater Seattle area that are suitable for stormwater detention basin construction. The four locations in total drain about 1200 ha of industrial area that releases approximately 0.6 kg TSS/ha/year.
- TSS removal by detention basins has been found to range from 60% to 90% (Minnesota Stormwater Manual 2021). Cost estimates include approximately $50,000 for construction of each detention basin, with a 10-year maintenance cycle that costs approximately $500 per basin.
- Seattle has identified 100 sites that could support the planting and maintenance of 0.1 ha rain gardens, with each rain garden draining 2 ha of residential area that generate approximately 0.2 kg TSS/ha/year.
- A literature review by Winer (2000) cited a TSS removal rate of 85% by rain gardens. Cost estimates are approximately $500 for initial construction of each rain garden and $100 for ongoing annual maintenance.

- The Seattle City Manager has proposed replacing all city parking lot surfaces (600 lots totaling 240 ha) with porous pavement on their next repaving cycle. The current parking lot surfaces typically discharge 0.4 kg TSS/ha/year.
- Replacing impervious surfaces with different types of porous pavements can remove between 92% and 98% of TSS (Brown et al. 2009). Each lot would cost $250 to repave and there would be no additional costs over a 10-year repaving cycle.

4.2.7 Build Your Foundational Knowledge

Below are web sources that provide additional information about each of the solutions you're considering for this Unit Challenge. This information can build a critical foundation to help you evaluate each option and support your final choice. After reviewing each source, be prepared to answer questions in the Preparation Assessment Quiz and to summarize any information relevant to your Unit Challenge.

Stormwater basins:
How Detention and Retention Ponds Work

Urban rain gardens:
Stormwater Best Management Practice. Bioretention (Rain Gardens)

Porous pavement:
Evaluating the potential benefits of permeable pavement on the quantity and quality of stormwater runoff

Final Product: A one-page Fact Sheet summarizing the issue, detailing your solution, and justifying your choice. Consider your audience, city planners who will use this Fact Sheet to help determine how to invest critical resources. Be sure to demonstrate how your proposed solution will stand up to the challenges posed by climate change.

4.2.8   Preparation Assessment Quiz

Are you ready to tackle your challenge? At this point you should understand the basic environmental principles and ecological processes involved in this complex environmental problem. Consider the following questions. If you are comfortable answering these, then you are ready to head into Discovery, Analysis, and Solutions activities.

- What have been the most important historic threats to the Puget Sound ecosystem?
- What physical features of the Sound make it so vulnerable to human activities?
- Why is Puget Sound able to support such diverse populations of marine creatures?
- Why is climate change such a difficult challenge for managers of the Puget Sound ecosystem?
- According to the Wessler Engineering article, what's the difference between detention and retention basins?
- According to the US EPA document, parking lots should have how much slope to be ideal for rain gardens?
- According to the USGS reference, what are the four potential benefits of using porous pavement?
- For each of your proposed solutions, are there any additional benefits that might arise from their implementation that might not be directly related to your Unit Challenge?
- For each of your proposed solutions, are there any negative unintended consequences that might result from their implementation?
- What additional information did you glean from your web sources that might help inform your Unit Challenge?

4.3   Puget Sound: Discovery

Specific Skills You'll Need to Review: Navigating the Scientific Literature, Science Communication, Problem-Solving

4.3.1   Independent Research

(Key Skill: Navigating the Scientific Literature)

To better understand the various approaches being considered to reduce inputs of TSS into the Puget Sound ecosys-tem from Seattle, you first need to examine the literature to see what others have found. Conduct a search of the peer-reviewed scientific literature focused on the solution you have been assigned and identify one research paper that focuses on your assigned approach.

Prepare a summary of the article you selected that includes the following:

- **Citation**
- **Main topic**: Stick to a few words, likely pulled from the title.
- **General summary**: A few bulleted sentences summarizing the research question it addresses and approaches it takes.
- **Methods:** How did they approach their research question?
- **Location**: Where was the work done?
- **Conclusions**: Concise list of the findings, specifically capturing the take-home message.
- **Relevance**: How might this study help inform your Unit Challenge? Feel free to make a bulleted list of information you may want to include later.

4.3.2   Literature Share: Reciprocal Instruction

(Key Skill: Scientific Communication)

Share In small groups, share and critique the research article you found. Keep in mind that your peers have not read this article, and it is your job to convey the key information to them. Note the items that will be important to consider when you are developing your solution to the challenge.

Critique Evaluate how these studies might help inform your Unit Challenge. Consider the following:

- Source (Quality of the work or bias of the authors)
- Methods (Did their methods sufficiently address the research question?)
- Conclusions (Did the results justify the conclusions made?)
- Relevance (Can these findings be applied to your challenge?)

Based on your critique, choose one article to share with the larger class, along with the key information that may be useful in deciding on a solution to propose.

4.3.3   Think-Compare-Share

(Key Skill: Problem–Solving)

Now that you have more information about possible solutions for this unit's challenge, you need to **develop a more formal problem definition** to guide your work throughout the rest of the exercises.

Think Start by working independently to develop a specific Problem Statement to guide the remainder of your work. Problem Statements provide the relevant information and boundaries to make the issue something you can effectively assess and tackle. The basics of a formal Problem Statement include the following:

Problem Statement: A short, concise statement summarizing the issue that includes the following:

- A **description** of the undesired condition or change that you hope to achieve (What is the actual problem?)
- **Justification** for addressing the problem (Why does this problem matter?)
- Potential **sources** or **causes** of this problem (What is the cause you need to address?)
- The **metrics** you will use to assess the status of the problem (How will you know if you are making a difference in the problem?)
- The **desired outcome** for these metrics (What is the end goal or ideal state?)
- Potential **solutions** to consider (How might you attempt to achieve this goal?)

Compare/Share Now return to your small group to share your Problem Statements. Use each of your ideas to develop a joint statement that contains all key information and is concise, clear and well written.

Unit Discovery Summary *Submit a final Problem Statement that succinctly captures the key information to guide your work on this Unit Challenge.*

4.3.4   Reflecting on your Work

(Key Skill: Personal Reflection)

After your work in Discovery, you should have a better idea of the problems you face and have produced a Problem Statement you can use to tackle the Unit Challenge. Take a moment to reflect on this work.

Consider the following prompts but feel free to expand on any to best capture your learning experience and better inform your next steps.

- Of the skills you practiced today, which were the most challenging? Which were the most interesting?
- How were you most comfortable working during these exercises? In small groups, independently, or with the larger class? Why? How does your choice reflect your personality type and leadership style?
- Was your Problem Statement strictly focused on the environmental problem of stormwater pollution, or did it also consider important social and economic considerations? How might a focus on the environmental aspects limit your ability to identify truly sustainable solutions?
- You've been given three viable solutions to assess as a part of this case study. But this is not an exhaustive list of options or even necessarily the best possible course of action for every scenario. Take a moment to "think outside the box." Are there any other possible solutions you think would be worth exploring? Describe one that you think would be worth pursuing.

4.4   Puget Sound: Analysis

Specific Skills You'll Need to Review: Quantitative Literacy, Sustainability Science

Review your Background and Discovery sections before beginning the Rotating Station exercises below. While you focused on one potential solution in your Independent Research in Discovery, keep an open mind as your work through Analysis activities.

4.4.1   Rotating Stations

(Key Skill: Quantitative Literacy)

At each of the following stations, you will review data that are relevant to the three potential solutions you're considering. Spend some time working through the analyses at each station to learn more about this issue and possible solutions for your Unit Challenge.

Be sure to write down one finding at each station that will help inform your selection of a solution.

Station 1: Stormwater Detention Basins Detention basins are commonly used to remove TSS from storm water and prevent its transport to surface waters. However, some nutrients carried in storm water may be harder to retain due to their soluble nature. A study by Soberg et al. (2020) compared TSS and phosphorus (P) removal by detention basins under various design and temperature regimes.

Figure 4.7 shows total phosphorus (TP) and TSS concentrations in detention basin outflows measured at two different temperatures (4.6 and 17.1 °C): at two different concentrations of salt, a commonly used additive to combat icy roads (a control and a 2000 mg Cl/L treatment), and at two different carbon treatments (a control and a 450 mm cellulose-based carbon layer (SZC) added).

After examining this figure, answer the following:

- How is the removal of TP and TSS impacted by salt in the water (0 vs. 2000 Salt box plots)?
- How is the removal of TP and TSS impacted by the addition of the SZC carbon layer (0 vs. 450 SZC box plots)?
- Which treatment (temperature, salt concentration, or carbon source) appears to have the greatest removal of TP? Why might that be?
- Which treatment (temperature, salt concentration, or carbon source) appears to have the greatest impact on TSS? Why might that be?

- Considering that climate change will result in warmer temperatures, what are the implications for TP and TSS removal by detention basins over time?

Record any relevant Station 1 findings

Station 2: Urban Rain Gardens In a recent study, Zhang et al. (2020) used a stormwater management model to estimate the ability of rain gardens to reduce outflows of stormwater runoff and retain TSS, total nitrogen (TN), and TP. Table 4.1 shows the percent runoff reduction and pollutant retention across a broad geographic range of rain gardens.

Based on these results, answer the following questions:

- Percent runoff retention across these studies varies widely. What might be the reason for this wide range of results?
- Using the median of any ranges presented, calculate the average retention for runoff, TSS, TN, and TP.
- Based on these averages, are rain gardens as effective at removing TN and TP as they are at removing TSS? Why might this be?

Fig. 4.7 Box plots of TP and TSS outflow concentrations for two temperatures, two salt concentrations, and with (right) and without (left) an embedded carbon source. The boxes indicate the interquartile range (IQR); the whiskers extend to 1.5 IQR, and outliers are indicated by an asterisk. Note the different scales of y-axes. (Source: Søberg et al. 2020 (Open Access))

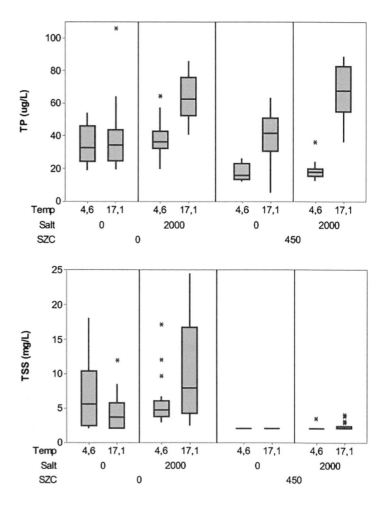

Table 4.1 Summary of percent runoff and pollutant retention by rain gardens across various studies

Location	Runoff	TSS	TN	TP
North Carolina, USA	78	–	40	65
Xi'an, China	97	–	–	–
Melbourne, Australia	33	–	–	–
Washington, USA	48–74	87–93	–	67–83
Istanbul, Turkey	23–85	–	–	–
Maryland, USA	–	22	–	74
North Carolina, USA	–	83	62	48
Maryland, USA	–	93	–	90
Indiana, USA	26	54	34	47
Kuala Lumpur, Malaysia	23	41	29	–

Source: Zhang et al. (2020 {Open Access})

- The TSS retention figure given in the Relevant Facts and Assumptions section above differs from the values in the table. Explain why this might have occurred. Which value/s would you use when choosing your solution and why?

Record any relevant Station 2 findings

Station 3: Porous Pavement Yu et al. (2021) examined the performance of permeable brick and other types of permeable pavement systems (PPS) in removing TSS and TP from runoff (Fig. 4.8). Contaminated water was continuously "rained" onto these various surfaces, followed by measurements of TSS and TP concentrations in outflows at various time intervals.

Examining the figure above, answer the following questions:

- Which PPS were best able to remove TSS? Were these same systems also able to effectively remove TP?
- How did the duration of rainfall inputs alter the performance of the PPS?
- Which PPS would you recommend? How might its performance change with the expected future climatic changes?

Record any relevant Station 3 findings

Station 4 The overall goal for this challenge is to remove TSS from stormwater runoff across the Seattle metropolitan area. In the description of your Unit Challenge, you were presented with several assumptions you can use to determine the total amount of TSS that may be removed by each solution, as well as the initial and long-term costs for maintenance. You will need this information to help inform your

decision-making later in Solutions, so be sure to save your calculations.

- Start by calculating the total kg TSS each approach would remove per year from the storm water based on the assumptions shown in Table 4.2.

Similarly, calculate the estimated 10-year costs for each approach (based on initial construction and maintenance costs):

	Rain garden	**Detention basin**	**Porous Pavement**
Land cover type	Urban residential	Urban industrial	Urban commercial
Number of treatments	100	4	600
Construction cost/ treatment ($)	500	50,000	250
Yearly maintenance/ treatment ($)	100	500	0
Total estimated 10-Year costs ($)			

Record any relevant Station 4 findings

4.4.2   System Mapping

(Key Skill: Sustainability Science)

Your assigned solution would have impacts beyond reducing the amount of TSS entering the Puget Sound ecosystem during wet weather. It may have a variety of direct and indirect economic, social, and ecological impacts (e.g., aesthetic, wildlife, groundwater recharge, landscape diversity, etc.) that may lead to additional positive or negative outcomes for a large urban ecosystem like Seattle.

Working with other class members assigned the same solution, develop a simple sustainability map that shows the various system connections across each of the three sustainability domains (ecological, economic, and societal). When you find connections between impacts that cross domains, draw a line to identify the connection (e.g., increase in property values with economic benefits leading to increased social stability).

The goal is to think broadly about this larger system and how implementing actions in one domain (e.g., ecological) may impact components in another domain (e.g., economic or societal).

Use the following template (Fig. 4.9) to get started:

When each group has completed their basic sustainability map, come together as a class to compare maps for the three

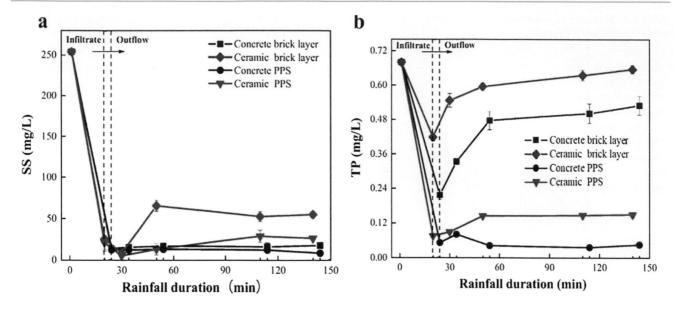

Fig. 4.8 (**a**) Total suspended solid (SS) and (**b**) total phosphorus (TP) concentrations before and after infiltration by the pavement. (Source: Yu et al. 2021)

Table 4.2 Assumptions for TSS removal by treatment type

	Rain garden	Detention basin	Porous Pavement
Land cover type	Urban residential	Urban industrial	Urban commercial
TSS released (kg/ha/year)	0.2	0.6	0.4
Treated area (ha)	200	1210	240
Percent TSS removal	85	75	95
Total kg TSS removal/year			

possible solutions. This information will help inform your decision support work later in the unit.

Unit Analysis Summary Based on your explorations, what have you learned that can help inform your choice of a solution? Do the data support the adoption of one of these potential solutions?

4.4.3   Reflecting on Your Work

(Key Skill: Personal Reflection)

During your work in Analysis, you explored some of the research into possible solutions to help inform your decision. Take a moment to reflect on this work. Consider the following prompts but feel free to expand on any to best capture your learning experience and better inform your next steps.

- How did you feel working with data? Do you consider quantitative literacy a strength or an area for improvement for you?
- How important should science be in informing management and policy? Do you feel the data you examined support and justify the costs of addressing your challenge?
- Any solution should be examined using a sustainability science lens. Your goal is to reduce the inflow of TSS carried from Seattle into the Puget Sound ecosystem. But any solution you implement may have other direct and indirect impacts. What are other possible economic, social, or ecological impacts? How should these considerations influence your decision?

4.5   Puget Sound: Solutions

Specific Skills You'll Need to Review: Problem-Solving, Decision Support, Communicating Science

Review your Unit Challenge and the major findings from Discovery and Analysis, including the sustainability map you created to highlight connections between possible solutions and the larger socioecological system.

4.5.1   Small Group Guided Worksheets

(Key Skill: Decision Support)

Decision support matrices can help break down the desired outcomes to reflect multiple criteria for consider-

Fig. 4.9 Template sustainability map to help connect ecological, societal, and economic considerations associated with your potential solutions

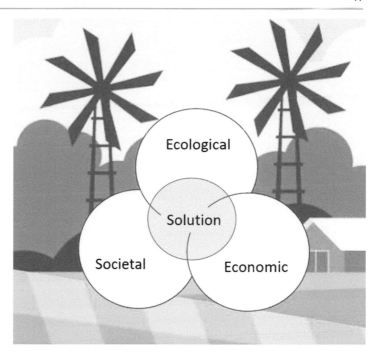

ation, and they allow you to compare how each possible solution achieves those desired outcomes. This not only helps inform decision-making, but it also provides transparency in the decision process and justification to help you advocate for its adoption.

For the three possible solutions, you will evaluate how well each can achieve the following desired outcomes:

- Reduces the amount of TSS entering Seattle's waterways from stormwater runoff
- Minimizes costs of installation and maintenance over a 10-year period
- Provides a lasting solution (not a short-term fix), even with a changing climate
- Is widely accepted by the community and raises awareness about water quality and climate change
- Minimizes secondary impacts (e.g., to wildlife or city aesthetics)

Considering your three potential solutions, develop a formal decision support matrix to compare and evaluate each approach using the template matrix below (Table 4.3). Getting the most out of the decision matrix requires a depth of knowledge about each of the three possible solutions. Below we list additional sources about each option. Please review each of these, paying particular attention to the two alternatives to your assigned approach.

Note that your group may have uncovered a novel solution not included in this list of three. You may choose to work through the structured decision matrix with your self-identified solution as a fourth solution option.

4.5.2 Additional Sources

1. Stormwater detention basins: Design Standards Chapter 1- General Information(iowadnr.gov)
2. Rain gardens What Is a Rain Garden and Why Is It Important? (thespruce.com)
3. Porous pavements: An Introduction to Porous Pavement | Home & Garden Information Center (clemson.edu)

Score each of your three solutions for each of the desired outcomes using a simple relative scale of 1 for least benefit to 5 for greatest benefit. Using a relative scale means you don't need to know exactly how well each solution meets the goal of each desired outcome, but you can use your judgment to assess how well each solution works compared to the others.

While this is a relative (subjective) scale, note that you will need to justify your scoring of each solution for each desired outcome.

Once each cell in your decision matrix has a relative score, calculate an average score for each solution. Based on this analysis, which is the "best" solution considering all your desired outcomes?

4.5.3 Role Playing

(Key Skill: Communicating Science)

In tackling environmental issues, you'll often find yourself working with groups of people with different perspectives about implementing a solution. You should use clear,

Table 4.3 Basic decision matrix to compare the three possible solutions for your Unit Challenge. Include your justification for each solution's ranking for each desired outcome

Solution Options	Desired Outcomes					
	Reduces TSS inputs to surface waters	Minimizes costs	Maximizes longevity of effect	Maximizes stakeholder buy-In	Minimizes secondary impacts	
Stormwater detention basins						
Urban rain gardens						
Porous pavement installations						
Justification for ranking						

Use a scale of 1-5 to score each option under each desired outcome category where 1 = does not achieve the desired outcome and 5 = completely achieves the desired outcome.

concise communication to summarize your solution and justify its selection. To be effective, this must address concerns likely to be presented by various stakeholder groups and how the risks of taking these actions outweigh the risks of taking no action.

Your group will present and justify your chosen solution to the class with a particular emphasis on how it might benefit or impact one of the key stakeholder groups listed below. Be sure to include arguments that support this solution from the ecological, societal, and economic domains of sustainability science. Listeners will also be **assigned a stakeholder identity**, with an opportunity to **ask follow-up questions** after your presentation that reflect their unique concerns and perspectives. Your job will be to listen carefully and tailor your answers to this audience of stakeholders.

Key stakeholder groups include the following:

- Property owners affected by the stormwater control installation
- Federal regulators charged with maintaining water quality in the Duwamish River
- A Native American member of the local chapter of Save the Chinook
- A representative from the Seattle Chamber of Commerce
- A member of the Beautify Seattle Committee

4.5.4   Reflecting on Your Work

(Key Skill: Personal Reflection)

During your work in Solutions, you have explored several possible actions that could reduce inputs of TSS into Seattle's waterways during storms. Take a little more time now to reflect on your findings and the skills you practiced. Consider the following prompts but feel free to expand on any of them to best capture your learning experience and feelings about this issue.

- Reflect on your work today through a sustainability science lens. Does your solution address ecological, economic, and societal considerations? Which considerations do you think should carry the most weight in the decision? Why?
- How did you weigh solutions that might have the greatest environmental impact against those that are most likely to be implemented and maintained for long-term impact?
- How can environmental scientists work to show the value of healthy ecosystems and justify the costs of mitigation strategies?
- Science communication can be challenging, especially when working with diverse audiences. We need to craft our communication to match the interests and values of

the target audience, but how do you do this when your audience contains a mix of stakeholder groups? How can you maximize the impact of your message to a diverse audience?

Unit Solutions Summary *Summarize and justify your final solution choice and outline how it addresses the direct challenge while also considering social, economic, and ecological impacts. Also demonstrate that it will continue to meet the challenges posed by climate change.*

4.6 Puget Sound: Final Challenge

As a part of this Unit Challenge, you were asked to write a one-page Fact Sheet justifying your choice of an approach to reduce the anticipated increase in TSS levels entering Puget Sound from Seattle as climate change progresses. Your Fact Sheet should include the following components:

- Brief problem statement
- Recommended mitigation strategy with sufficient details to summarize the general approach
- Justification of this recommendation (e.g., long-term effectiveness given anticipated climate changes, implementation costs, other benefits provided, etc.). Be sure to use a sustainability lens to include considerations of direct and indirect ecological, social, and economic considerations
- Any obstacles the group might face trying to implement your solution

Consider using figures, graphics, and tables to help summarize the system and show how this solution is well suited to meet all desired outcomes.

Final Unit Challenge Submit your final recommendations in a one-page Fact Sheet using clear science communication designed for a lay audience.

References

Bash J, Berman CH, Bolton S (2001) Effects of turbidity and suspended solids on salmonids. Wash State Trans Cent Final Res Rept Res Proj T1803.Task 42

Brown C, Chu A, van Duin B, Valeo C (2009) Characteristics of sediment removal in two types of permeable pavement. Water Qual Res J Can 44(1):59–70

Mauger GS, Casola JH, Morgan HA, Strauch RL, Jones B, Curry B, Busch Isaksen TM, Whitely Binder L, Krosby MB, Snover AK (2015) State of knowledge: climate change in Puget Sound. Report prepared for the Puget Sound Partnership and the National Oceanic and Atmospheric Administration. Climate Impacts Group, U. Wash., Seattle. http://hdl.handle.net/1773/34347

Minnesota Stormwater Manual (2021) Total suspended solids (TSS) in stormwater-Minnesota Stormwater Manual. stormwater.pca.state.mn.us

Søberg LC, Al-Rubaei AM, Viklander M, Blecken G-T (2020) Phosphorus and TSS removal by stormwater bioretention: effects of temperature, salt, and a submerged zone and their interactions. Water Air Soil Pollut 231(6):270. https://doi.org/10.1007/s11270-020-04646-3

The Puget Sound Water Quality Authority (1986) The state of the Sound 1986. Prepared with assistance of Entranco Eng. Inc. 198 p

Winer R (2000) National pollutant removal performance database for stormwater treatment practices, 2nd edn. US EPA Office of Science and Technology. 29 p

Yu Z, Gan H, Xiao M, Huang B, Zhu DZ, Shang Z (2021) Performance of permeable pavement systems on stormwater permeability and pollutant removal. Environ Sci Pollut Res 28:28571–28584

Zhang L, Ye Z, Shibata S (2020) Assessment of rain garden effects for the management of urban storm runoff in Japan. Sustainability 12(23). https://doi.org/10.3390/su12239982

Zier J, Gaydos JK (2016) The growing number of species of concern in the Salish Sea suggests ecosystem decay is outpacing recovery. In: Proceedings of the 2016 Salish Sea Ecosystem Conference. Vancouver, pp 1–17

Core Knowledge
Biogeochemical cycles, Agriculture, Soils, Water quality, Resource management

5.1 Environmental Issue

Movement of excessive amounts of nitrogen (N) through the Mississippi River Basin (MRB) into the Gulf of Mexico (GOM) (Fig. 5.1) has been linked to the annual formation of a Dead Zone, an area of reduced dissolved oxygen (DO) levels in the waters around the river's mouth. Dead Zone formation has been linked primarily to agricultural practices in the MRB, and altered conditions associated with climate change will likely affect both the movement of N into the GOM and the frequency and extent of conditions that favor the formation of the Dead Zone.

Understanding how the changing climate will affect nutrient cycling and conditions in the GOM will be critical if we are to effectively mitigate the impacts of Dead Zone formation. In this unit, you'll look at the big picture of the Gulf's Dead Zone and then consider what an individual farmer can do about the problem.

5.2 Background Information

5.2.1 The Problem

The GOM suffers from hypoxia (water containing < 2 mg/L DO) that occurs most frequently during warm months. While there are over 400 areas of hypoxia or Dead Zones in the world (up from six in the 1960s), the hypoxic zone in the GOM is one of the largest. The worst hypoxia event to date in the Gulf occurred in the summer of 2017, covering 22,729 km², an area about the size of New Jersey.

The impact of Dead Zone formation on the Gulf's ecosystem can be devastating. Direct effects include fish kills, which deplete valuable fisheries and disrupt ecosystems. Mobile animals (e.g., adult fish) can typically survive a hypoxic event by moving to waters containing more DO. Less mobile or immobile animals, such as mussels or crabs, are often killed during hypoxic events. Ultimately, Dead Zones can cause a severe decrease in the amount and diversity of life in affected areas.

It's important to stress that, in addition to the environmental impacts that result from N movement into the GOM, the economic toll resulting from hypoxic conditions, including the loss of commercial fisheries and tourism, can also be devastating.

More than half of all pesticides and fertilizers used in the United States are applied within the MRB. Draining more than 3.2 million km², this watershed is the third largest and one of the most productive river basins in the world, with about 58% of its area in cropland (Goolsby and Battaglin 2000). While scientists have recently suggested a role for phosphorus in Dead Zone formation, nitrogen (N) has been pinpointed as the main trigger for the annual formation of this phenomenon.

Fertilizers have been identified as being particularly important (Fig. 5.2), with 52% of the N moving into the Gulf estimated to originate from the use of fertilizers on corn and soybean crops alone (Alexander et al. 2014).

Surges in stormwater runoff following spring rains often coincide with the timing of fertilizer application. N contained in runoff that eventually enters coastal waters like the GOM can stimulate an overgrowth of algae (phytoplankton) in these waters. The algae eventually die and sink to deeper waters of the Gulf, where they are decomposed by bacteria, using up oxygen. Under the right conditions, the bottom waters become severely depleted of DO, stressing fish and suffocating other marine life that can't escape (Fig. 5.3).

J. Pontius, A. McIntosh, *Environmental Problem Solving in an Age of Climate Change*, Springer Textbooks in Earth Sciences, Geography and Environment, https://doi.org/10.1007/978-3-031-48762-0_5

Fig. 5.1 This figure shows how runoff from farms (green areas) and cities (red areas) drains into the Mississippi River, delivering nutrients into the Gulf of Mexico and fueling the annual hypoxic zone formation. (Source: NOAA {Public Domain})

Fig. 5.2 Estimated nitrogen contribution to the Gulf of Mexico from specific sources. (Source: USGS {Public Domain})

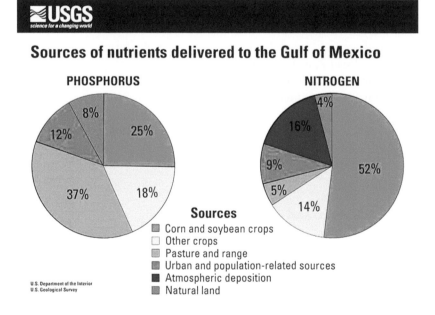

5.2.2   The Role of Climate Change

Scientists believe that projected climatic changes will worsen the prevalence and impacts of Dead Zones. Warmer waters hold less DO, making it easier for Dead Zones to form at the same time the metabolism and oxygen demands of marine organisms increase. Changes in the frequency and intensity of precipitation events are also linked to increased N runoff. The size of the Dead Zone in any given summer will be significantly influenced by spring weather conditions—other things being equal, the wetter the spring in the watershed, the greater the nutrient pollution, the bigger the algal blooms, and the larger the Dead Zone.

Fig. 5.3 The formation of the Dead Zone. (Source: Dan Swenson (CC BY 2.0) via Flickr)

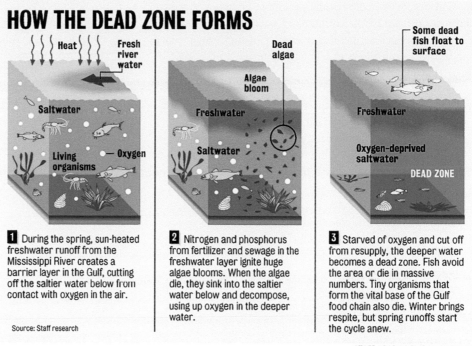

HOW THE DEAD ZONE FORMS

1 During the spring, sun-heated freshwater runoff from the Mississippi River creates a barrier layer in the Gulf, cutting off the saltier water below from contact with oxygen in the air.

2 Nitrogen and phosphorus from fertilizer and sewage in the freshwater layer ignite huge algae blooms. When the algae die, they sink into the saltier water below and decompose, using up oxygen in the deeper water.

3 Starved of oxygen and cut off from resupply, the deeper water becomes a dead zone. Fish avoid the area or die in massive numbers. Tiny organisms that form the vital base of the Gulf food chain also die. Winter brings respite, but spring runoffs start the cycle anew.

Source: Staff research

STAFF GRAPHIC BY DAN SWENSON

The process is further driven by the indirect effects of increased atmospheric CO_2 levels on crop production and warmer temperatures on biogeochemical processes. As atmospheric CO_2 levels continue to rise, some crops may increase their productivity, resulting in an increased need for fertilizer application. Warmer temperatures will also increase nitrification rates and the conversion of ammonia to more mobile nitrate compounds that can be transported into surface waters. Increased precipitation projected to occur in the MRB will also lead to increased N loss from soils through leaching.

5.2.3 Solutions

There are several approaches to help minimize the formation of the GOM Dead Zone: (a) controlling N at its source within the MRB, (b) removing N from the river itself, or (c) minimizing impacts once N reaches the GOM. Of these three, the most effective solution likely lies at the source: reduce the movement of N into the system in the first place.

It's hard to overstate the enormity of the challenge posed by nutrient pollution in the MRB. The basin covers nearly 40% of the lower 48 US states. In states like Iowa, where 70% of the land is intensively farmed, the amount of N discharged into the Mississippi River is immense, totaling more than 500,000 tons as nitrate in 2017 (Newman et al. 2019).

Despite the challenges posed within the MRB, scientists and policymakers have proposed a number of possible solutions. For instance, to tackle this problem at its source, the Gulf of Mexico Hypoxia Task Force has evaluated a wide range of strategies to reduce nutrient pollution, including market-based incentives for wetland restoration, home septic system improvements, planting of cover crops, and better stormwater management. Scientists and economists have analyzed various options and estimated the costs to plant more cover crops, install drainage filters, and change land use to be about $2.7 billion a year basin-wide (Rabotyagov et al. 2014).

5.2.4 Unit Challenge

The overarching problem we face in addressing the GOM's Dead Zone formation is how to reduce agricultural inputs of nutrients into waters upstream while accounting for the ongoing and potential impacts of climate change on the function of the larger system. Your Unit Challenge is **to evaluate N reduction strategies at the individual farm level.**

5.2.5 The Scenario

A family farm has 200 hectares (ha) in corn and soybeans alongside Iowa's Cedar River, a tributary to the Iowa River which drains into the Mississippi River. As a committed environmentalist, the farmer recognizes the need to reduce the surface runoff and subsurface drainage of N. Reducing fertilizer use is not an option, and she needs you, an expert in agricultural ecosystems working for the US Department of

Agriculture, to evaluate several general approaches to reduce N transport from her fields into the Cedar River.

After considering the characteristics of the farmer's land and the various options to reduce N transport, you settle on three possible options:

- **Plant buffer strips.** Buffer strips are vegetated areas placed between actively farmed fields and surface waters. While the benefits of vegetated buffers vary depending on buffer width and vegetation type and age, research shows that vegetated buffers can intercept and retain greater than 90% of nitrate N and 50% of the P carried in runoff. Buffers can also provide important wildlife habitat, conserve soils, and mitigate floods.
  Specific recommendation: Design and plant 20 m wide vegetated buffers along the 1 km extent of Cedar River bordering this property.
- **Recycle drainage water**. In drainage water recycling, water running off agricultural fields is captured in a nearby detention basin and then used to irrigate crops during dry periods. In addition to helping meet crop water demands, drainage water recycling also returns nutrients like N and P in runoff back to the agricultural fields where they can increase crop productivity.
- *Specific recommendation*: Design and install a drainage water recycling system to capture and reuse runoff from the 200 ha of actively tilled land on this property.
- **Establish constructed wetlands**. Constructed wetlands are designed to replicate the ability of natural wetlands to remove pollutants before they are discharged to nearby surface waters. The ecological design of constructed wetlands uses plants and microbes to remove and retain up to 98% of N, P, and total suspended solids (TSS) contained in runoff. Engineered properly, constructed wetlands can even be used to treat sewage.
- *Specific recommendation*: Design and install a 10 ha constructed wetland to capture and treat water running off the 200 ha of actively farmed land on this property.

You will discover that each of these options has various pros and cons and different costs and benefits. Your job is to evaluate these options and recommend an approach to the farmer that will perform best over time under the conditions expected with climate change.

5.2.6 Relevant Facts and Assumptions

- Size of the actively farmed land is 200 ha, with 1 km of stream bank bordering the cropped area.

- Estimated costs associated with each approach:
  - Buffer strip creation: To install a 20 m-wide buffer strip along the full length of the 1 km stream costs $5.50/m^2. Additional crop yield production losses are $1000/ha.
  - Drainage water recycling system: $10,000 for a 10-ha detention basin construction; $200/ha for drainage diversion; $350/ha irrigation system installation. Additional crop yield production losses are $1000/ha.
  - Constructed wetland: $40,000 for construction of a 10-ha wetland; $200/ha for drainage diversion. Additional crop yield production losses are $1000/ha.

- Funding is limited to $150,000 in conservation grants.
- Soil type is prairie-derived till and loess with high organic matter content and moderate permeability.
- 100 kg of fertilizer are applied per ha annually.

5.2.7 Build Your Foundational Knowledge

Below are web sources that provide additional information about each of the solutions you're considering for this Unit Challenge. This information can build a critical foundation to help you evaluate each option and support your final choice. After reviewing each source, be prepared to answer questions in the Preparation Assessment Quiz and to summarize any information relevant to your Unit Challenge.

Buffer strips:
The multifunctional roles of vegetated strips around and within agricultural fields

Drainage water recycling:
Questions and Answers about Drainage Water Recycling for the Midwest

Constructed wetlands:
Constructed Treatment Wetlands

Final Product: A one-page Fact Sheet summarizing the issue, detailing your solution, and justifying your choice of that solution. Consider your audience, the farm manager, and the conservation organization that provides funding for the project. Be sure to demonstrate how your proposed solution will stand up to the challenges posed by climate change.

5.2.8 Preparation Assessment Quiz

Are you ready to tackle your challenge? At this point you should understand the basic environmental principles and ecological processes involved in this environmental problem. Consider the following questions. If you are comfortable with answering these, then you are ready to head into Discovery, Analysis, and Solutions activities.

- What is hypoxia in aquatic systems and what causes this condition?
- Why is it N and not P that most limits primary production in the Gulf?
- Which form of N is most mobile in ecosystems?
- How might warmer temperatures and heavier precipitation expected with climate change impact biogeochemical processes such as the N cycle? Dead Zone formation?
- The article by Haddaway et al. identified a variety of services provided by vegetated buffer strips around agricultural fields. Identify any two of these.
- Frankenberger et al. name two major benefits of drainage water recycling. What are they?
- The US EPA Fact Sheet on constructed wetlands identifies improved water quality as a benefit but also names an additional benefit. What is it?
- For each of your proposed solutions, are there any additional benefits that might arise from their implementation that might not be directly related to your Unit Challenge?
- For each of your proposed solutions, are there any negative unintended consequences that might result from their implementation?
- What additional information did you glean from your web sources that might help inform your Unit Challenge?

5.3 Gulf of Mexico: Discovery

Specific Skills You'll Need to Review: Navigating the Scientific Literature, Science Communication, Problem-Solving

5.3.1 Independent Research

(Key Skill: Navigating the Scientific Literature)

In order to better understand nitrogen (N) behavior in agricultural landscapes and the effectiveness of various mitigation strategies, you first need to examine the literature to see what others have found. Conduct a search of the peer-reviewed scientific literature focused on the solution you have been assigned and identify one research paper that focuses on your assigned approach.

Prepare a summary of the article you selected that includes the following:

- **Citation**
- **Main topic**: Stick to a few words, likely pulled from the title.
- **General summary**: A few bulleted sentences summarizing the research question it addresses and approaches it takes.
- **Methods:** How did they approach their research question?
- **Location**: Where was the work done?
- **Conclusions**: Concise list of the findings, specifically capturing the take-home message.
- **Relevance**: How might this study help inform your Unit Challenge? Feel free to make a bulleted list of information you may want to include later.

5.3.2 Literature Share: Reciprocal Instruction

(Key Skill: Scientific Communication)

Share In small groups, share and critique the research article you found. Keep in mind that your peers have not read this article, and it is your job to convey the key information to them. Note the items that will be important to consider when you are designing your solution to the challenge.

Critique Evaluate how these studies might help inform your Unit Challenge. Consider the following:

- Source (Quality of the work or bias of the authors)
- Methods (Did their methods sufficiently address the research question?)
- Conclusions (Do the results justify the conclusions made?)
- Relevance (Can these findings be applied to your challenge?)

Based on your critiques, choose one article to share with the larger class, along with the key information that may be useful in deciding on a solution to propose.

5.3.3 Think-Compare-Share

(Key Skill: Problem-Solving)

Now that you have more information about possible solutions for this unit's challenge, you need to **develop a more formal problem definition** to guide your work throughout the rest of the exercises.

Think Start by working independently to develop a specific Problem Statement to guide the remainder of your work. Problem Statements provide the relevant information and boundaries to make the issue something you can effectively assess and tackle. The basics of a formal Problem Statement include the following:

Problem Statement: A short, concise statement summarizing the issue that includes the following:

- A **description** of the undesired condition or change that you hope to achieve (What is the actual problem?)
- **Justification** for addressing the problem (Why does this problem matter?)
- Potential **sources** or **causes** of this problem (What is the cause you need to address?)
- The **metrics** you will use to assess the status of the problem (How will you know if you are making a difference in the problem?)
- The **desired outcome** for these metrics (What is the end goal or ideal state?)
- Potential **solutions** to consider (How might you attempt to achieve this goal?)

Compare/Share Now return to your small group to share your Problem Statements. Use each of your ideas to develop a joint statement that contains all key information and is concise, clear, and well written.

Unit Discovery Summary *Submit a final Problem Statement that succinctly captures the key information to guide your work on this Unit Challenge.*

5.3.4 Reflecting on Your Work

(Key Skill: Personal Reflection)

After your work in Discovery, you should have a better idea of the problems you face and have produced a Problem Statement you can use to tackle the Unit Challenge. Take a moment to reflect on this work. Consider the following prompts but feel free to expand on any to best capture your learning experience and better inform your next steps.

- Of the skills you practiced in Discovery, which were the most challenging? Which were the most interesting?
- How were you most comfortable working during these exercises? In small groups, independently, or with the larger class? Why? How does your choice reflect your personality type and leadership style?
- Was your Problem Statement strictly focused on the environmental problem of nutrient pollution, or did it also consider important social and economic considerations? How might a focus on the environmental aspects limit your ability to identify truly sustainable solutions?
- You've been given three viable solutions to assess as a part of this case study. But this is not an exhaustive list of options or even necessarily the best possible course of action for every scenario. Take a moment to "think outside the box." Are there any other possible solutions you think would be worth exploring? Describe one that you think would be worth pursuing.

5.4 Gulf of Mexico: Analysis

Specific Skills You'll Need to Review: Quantitative Literacy, Sustainability Science

Review your Background and Discovery sections before beginning the Rotating Station exercises below. While you focused on one potential solution in your Independent Research in Discovery, keep an open mind as your work through Analysis activities.

5.4.1 Rotating Stations

(Key Skill: Quantitative Literacy)

At each of the following stations, you will review data that are relevant to the three potential solutions you're considering. Spend some time working through the analyses at each station to learn more about this issue and possible solutions for your Unit Challenge.

Be sure to write down one finding at each station that will help inform your selection of a solution.

Station 1: Buffer Strips The data shown in Fig. 5.4 are from a comparison of buffer strip type and width around the eutrophic Lake Chaohu in China (Cao et al. 2018). Each buffer was examined for total N (TN) removal from stormwater runoff. The goal of this study was to identify the optimal width for buffer strips around this eutrophic lake to help remove TN from runoff with as little impact on the surrounding active agriculture as possible.

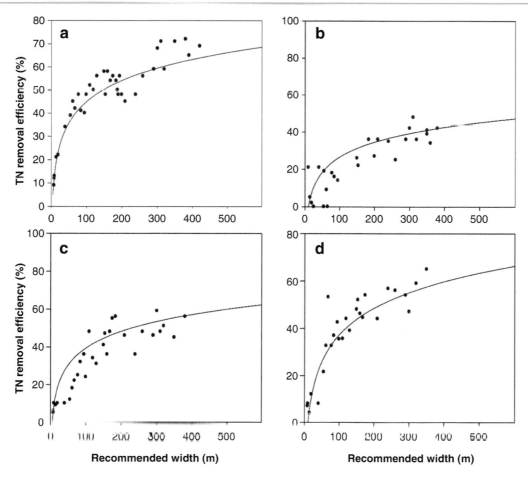

Fig. 5.4 Relationships between TN removal efficiency from storm water and the width of different types of riparian buffer in the Lake Chaohu watershed. (**a**) Grass; (**b**) Forest; (**c**) Wetland; (**d**) Grass/Forest. (Source: Cao et al. 2018 {Open Access})

Answer the following based on the graphs shown above:

- Is the relationship between buffer width and effective N removal linear or does the effectiveness begin to level off with greater widths? What does this tell you about the ideal width for a buffer strip to maximize TN removal?
- Which of the various vegetation types used for the buffer was best at removing TN overall? Why might the different vegetation types have had such different TN removal rates?
- If a farmer were willing to convert only a 10 m strip of field for use as a riparian buffer, which vegetation type should she choose and how effective would it be at removing TN? What width would you suggest this farmer adopt to double that TN removal efficiency?

Record any relevant Station 1 findings

Station 2: Constructed Wetlands Wetlands constructed to treat wastewater have also been used effectively to remove N from non-point sources like stormwater runoff (Table 5.1). Beutel et al. (2009) assessed how effective constructed wet-

Table 5.1 Summary of nitrate removal in the constructed treatment wetlands. Units include an area-based first-order removal rate constant for nitrate (K_N) and the rate normalized to 20 °C (K_{N-20})

Nitrate Concentration (mg-N/L)		K_N (m/year)	K_{N-20} (m/year)	Removal efficiency (%)	Removal rate (mg-N/m²/d)
In	Out				
Sedimentation basin					
2.0 (0.86)	1.4 (0.74)	187 (163)	237 (183)	34 (27)	837 (884)
North wetland					
1.4 (0.73)	0.1 (0.32)	142 (54)	196 (80)	93 (14)	139 (75)
South wetland					
1.3 (0.74)	0.2 (0.32)	149 (68)	192 (84)	90 (15)	146 (83)

Source: Beutel et al. (2009)
Values are averages and standard deviations (in parentheses) for 2003–2006 data set; n – 30–35

lands were at treating agricultural runoff from drainage tiles over several years in the Yakima Basin, WA.

Consider the data in the table and answer the following:

- Based on the findings in this table, how much more effective (removal efficiency) were their two test wetlands at removing N than the simple sedimentation basin?
- If a farmer's fields discharged 0.9 mg-N/L, how much could be initially removed by a simple sedimentation basin?
- If the discharge from this sedimentation basin was then passed through a system like the South wetland, what would the final concentration of N in the water be as it left the treatment system?

Record any relevant Station 2 findings

Station 3: Drainage Water Recycling Drainage water recycling involves capturing and storing water running off from agricultural fields and using it to irrigate crops. Reinhart et al. (2019) measured nitrate-N reduction in drainage water for various reservoir sizes at two tile-drained research farms in the US Midwest (DPAC and SERF) (Fig. 5.5). The percent shown on the X-axes refers to reservoir size as a percentage of the field area.

Based on the data shown in the two figures above, answer the following:

- Describe the relationship between the size of the reservoir and the percent nitrate reduction.
- Considering that larger reservoirs cost more to construct and take more land out of active planting, what size reservoir would you suggest installing? Why?
- Compare the two sites (DPAC and SERF). How does the percent reduction compare between the two locations?
- Assume that when the lower MPR standard deviation line crosses the Y intersect (y = 0), it means that there is no *widespread* reduction that results from the drainage recycling treatment. Do you believe this approach works to effectively reduce nitrate loads at both locations (DPAC and SERF)? What other factors might account for any differences?

Record any relevant Station 3 findings

Station 4 Earlier you were given a set of assumptions that included estimated costs for implementing each of your three possible solutions. Calculate the following for each approach:

- What are the initial installation/construction costs (year 1) for each of the three solutions?
- What are the 5-year installation/construction and crop loss costs for each of the three?

Fig. 5.5 The effectiveness of drainage water recycling based on reservoir size. (Source: Reinhart et al. 2019)

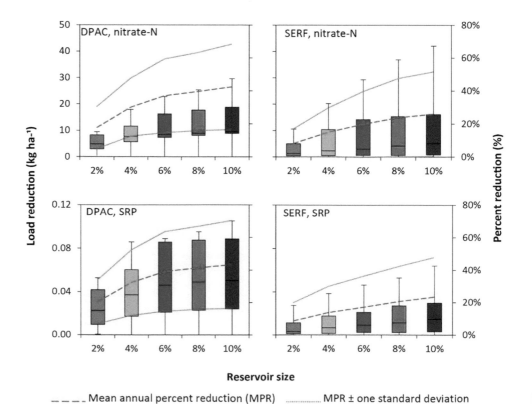

- Do you see a considerable difference in the costs for each potential solution? How much should these costs weigh in the selection of any one potential solution?

Record any relevant Station 4 findings

5.4.2 System Mapping

(Key Skill: Sustainability Science)

Your assigned solution would require some physical structure that will have impacts beyond their targeted reductions in N entering surface waters. It may have a variety of direct and indirect economic, social, and ecological impacts (e.g., aesthetic, wildlife, groundwater recharge, landscape diversity, etc.) that may lead to additional positive or negative outcomes for the larger socioecological system of the farm.

Working with other class members assigned the same solution, develop a simple sustainability map that shows the various system connections across each of the three sustainability domains (ecological, economic, and societal). When you find connections between impacts that cross domains, draw a line to identify the connection (e.g., enhanced wildlife habitat (ecological) leading to increased farm property value (economic)).

The goal is to think broadly about this larger system and how implementing actions in one domain (e.g., ecological) may impact components in another domain (e.g., economic or societal).

Use the following template (Fig. 5.6) to get started:

When each group has completed their basic systems map, **come together as a class to compare maps for the three possible solutions.** This information will help inform your decision support work later in the unit.

Unit Analysis Summary *Based on your explorations, what have you learned that can help inform your choice of a solution? Do the data support the adoption of one of these potential solutions?*

5.4.3 Reflecting on Your Work

During your work in Analysis, you explored some of the research into possible solutions to help inform your decision. Take a moment to reflect on this work. Consider the following prompts but feel free to expand on any to best capture your learning experience and better inform your next steps.

- How did you feel working with data? Do you consider quantitative literacy a strength or an area for improvement for you?
- How important should science be in informing management and policy? Do you feel the data you examined support and justify the costs of addressing your challenge?
- Any solution should be examined using a sustainability science lens. Your goal is to reduce N inputs from this farm to the Cedar River. But any solution you implement may have other direct and indirect impacts. What are other possible economic, social, or ecological impacts? How should these considerations influence your decision?

Fig. 5.6 Template sustainability map to help connect ecological, societal, and economic considerations associated with your potential solutions

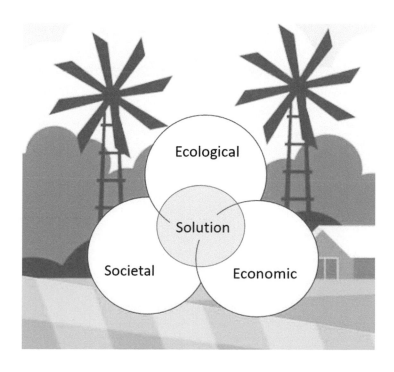

5.5 Gulf of Mexico: Solutions

Specific Skills You'll Need to Review: Problem-Solving, Decision Support, Communicating Science

Review your Unit Challenge and the major findings from Discovery and Analysis, including the sustainability map you created to highlight connections among possible solutions and the larger socioecological system.

5.5.1 Small Group Guided Worksheets

(Key Skill: Decision Support)

Decision support matrices can help break down the desired outcomes to reflect multiple criteria for consideration, and they allow you to compare how each possible solution achieves those desired outcomes. This not only helps inform decision-making, it also provides transparency in the decision process and justification to help you advocate for its adoption.

For the three possible solutions, you will evaluate how well each can achieve the following desired outcomes:

- Limits the amount of N entering surface waters
- Minimizes costs of installation and maintenance
- Provides a lasting impact (not a short-term fix), even with a changing climate
- Is widely accepted by the farmer and the community
- Minimizes secondary impacts (e.g., on wildlife or farm productivity)

Considering your three potential solutions, develop a formal decision support matrix to compare and evaluate each approach using the template matrix below (Table 5.2). Getting the most out of the decision matrix requires a depth of knowledge about each of the three possible solutions. Below we list additional sources about each option. Please review each of these, paying particular attention to the two alternatives to your assigned approach.

Note that your group may have uncovered a novel solution not included in this list of three. You may choose to work through the structured decision matrix with your self-identified solution as a fourth solution option.

Table 5.2 Basic decision matrix to compare the three possible solutions for your Unit Challenge. Include your justification for each solution's ranking for each desired outcome

Solution Options	Reduces N inputs to surface water	Minimizes installation costs	Maximizes longevity of effect	Maximizes stakeholder buy-in	Minimizes secondary impacts
Desired Outcomes					
Buffer strips					
Drainage recycling					
Constructed wetlands					
Justification for Ranking					

Use a scale of 1-5 to score each option under each desired outcome category where 1 = does not achieve the desired outcome and 5 = completely achieves the desired outcome.

5.5.2 Additional Sources

1. Buffer strips: Vegetation Buffer Strips in Agricultural Areas (chisagocountymn.gov)
2. Constructed wetlands: (PDF) The Use of Constructed Wetlands to Mitigate Pollution from Agricultural Runoff (researchgate.net)
3. Drainage water recycling: Drainage Water Recycling – Transforming Drainage

Score each of your three solutions for each of the desired outcomes using a simple relative scale of 1 for least benefit to 5 for greatest benefit. Using a relative scale means you don't need to know exactly how well each solution meets the goal of each desired outcome, but you can use your judgment to assess how well each solution works compared to the others.

While this is a relative (subjective) scale, note that you will need to justify your scoring of each solution for each desired outcome.

Once each cell in your decision matrix has a relative score, calculate an average score for each solution. Based on this analysis, which is the "best" solution considering all your desired outcomes?

5.5.3 Role Playing

(Key Skill: Communicating Science)

In tackling environmental issues, you'll often find yourself working with groups of people with different perspectives about implementing a solution. You should use clear, concise communication to summarize your solution and justify its selection. To be effective, this must address concerns likely to be presented by various stakeholder groups and how the risks of taking these actions outweigh the risks of taking no action.

Your group will present and justify your chosen solution to the class with a particular emphasis on how it might benefit or impact one of the key stakeholder groups listed below. Be sure to include arguments that support this solution from the ecological, societal, and economic domains of sustainability science. Listeners will also be **assigned a stakeholder identity**, with an opportunity to **ask follow-up questions** after your presentation that reflect their unique concerns and perspectives. Your job will be to listen carefully and tailor your answers to this audience of stakeholders.

Key stakeholder groups include the following:

- Farmers wanting to reduce N pollution generated by their lands
- A State of Iowa government water quality expert

- A member of the local chapter of Friends of the Cedar River
- A representative from the Association of Fertilizer Manufacturers
- A consumer group lobbying for lower corn prices

5.5.4 Reflecting on Your Work

(Key Skill: Personal Reflection)

During your work in Solutions, you explored several possible actions that could be taken to mitigate inputs of N pollution from a farmer's field into a nearby river. Take a little more time now to reflect on your findings and the skills you practiced. Consider the following prompts but feel free to expand on any of them to best capture your learning experience and feelings about this issue.

- Reflect on your work today through a sustainability science lens. Does your solution address ecological, economic, and societal considerations? Which considerations do you think should carry the most weight in the decision? Why?
- How did you weigh solutions that might have the greatest environmental impact against those that are most likely to be implemented and maintained for long-term impact?
- How can environmental scientists work to show the value of healthy ecosystems and justify the costs of mitigation strategies?
- Science communication can be challenging, especially when working with diverse audiences. We need to craft our communication to match the interests and values of the target audience, but how do you do this when your audience contains a mix of stakeholder groups? How can you maximize the impact of your message to a diverse audience?

Unit Solution Summary Summarize and justify your final solution choice and outline how it addresses the direct challenge while also considering social, economic, and ecological impacts. Also demonstrate that it will continue to meet the challenges posed by climate change.

5.5.5 Gulf of Mexico: Final Challenge

As a part of this Unit Challenge, you were asked to write a one-page Fact Sheet justifying your choice for an approach to reduce N inputs from a farm into a nearby river. Your Fact Sheet should include the following components:

- Brief problem statement

- Recommended mitigation strategy with sufficient details to summarize the general approach
- Justification of this recommendation (e.g., long-term effectiveness given anticipated climate changes, implementation costs, other benefits provided, etc.). Be sure to use a sustainability lens to include considerations of direct and indirect ecological, social, and economic considerations
- Any obstacles the group might face trying to implement your solution

Consider using figures, graphics, and tables to help summarize the system and show how this solution is well suited to meet all desired outcomes.

Final Unit Challenge Submit your final recommendations in a one-page Fact Sheet using clear science communication designed for a lay audience.

References

Alexander RB, Smith BA, Schwarz GE, Boyer EW, Nolan JV, Brakebill JW (2014) Differences in phosphorus and nitrogen delivery to the Gulf of Mexico from the Mississippi River Basin. Environ Sci Technol 42:822–830

Beutel MW, Newton CD, Brouillard ES, Watts RJ (2009) Nitrate removal in surfaceflow constructed wetlands treating dilute agricultural runoff in the lower Yakima Basin, Washington. Ecol Eng 35(10):1538–1546

Cao X, Song C, Xiao J, Zhou Y (2018) The optimal width and mechanism of riparian buffers for storm water nutrient removal in the Chinese eutrophic Lake Chaohu watershed. Water 10(10):1489

Goolsby DA, Battaglin WA (2000) Nitrogen in the Mississippi River Basin – estimating sources and predicting flux to the Gulf of Mexico. USGS Fact Sheet 135-00. https://doi.org/10.4144/fs13500

Newman J, Rigdon R, McGroarty P (2019) The world's appetite is threatening the Mississippi River. Wall Street J. http://graphics.wsj.com/mississippi/

Rabotyagov SS, Kling CL, Gassman PW, Rabalais NN, Turner RE (2014) The economics of dead zones: causes, impacts, policy challenges, and a model of the Gulf of Mexico hypoxic zone. Rev Environ Econ Policy 8(1):58–79

Reinhart BD, Frankenberger JR, Hay CH, Helmers MJ (2019) Simulated water quality and irrigation benefits from drainage water recycling at two tile-drained sites in the US Midwest. Agric Water Manag 223:105699

Core Knowledge

Climate change, Energy policy, Carbon footprints, Sustainability

6.1 Environmental Issue

Despite international agreements like the Paris Accord signed in 2015 by 196 nations which agreed to take measures to limit CO_2 emissions in order to keep global temperature increases below 2 °C, too little progress has been made towards meeting those goals (Fig. 6.1). Even the host nation France was found guilty of failing to meet the climate change goals it committed to in a 2021 court ruling. As if to underscore the collective lack of action to reduce carbon emissions, the highest ever recorded atmospheric CO_2 concentration, 423 ppm, was measured at Mauna Loa, Hawaii, in June of 2022.

In a nearly perfect correlation with changes in atmospheric CO_2 concentrations (Fig. 6.2), the Earth's average surface temperature has risen about 1 °C over the past 50 years, with impacts on weather patterns, extreme weather events, ecosystem function, and human societies already occurring. Such impacts are projected to become more widespread and severe if the Earth warms another 2 °C as predicted under a low-emissions scenario. If a 3 °C increase occurs, as projected by a high-emissions scenario, many scientists believe the Earth will reach a "tipping point," with a catastrophic collapse of many ecosystems and cascading impacts across human communities. Which of these projected climate outcomes we pass on to the next generation depends on our efforts now to reduce greenhouse gas (GHG) emissions globally.

In this unit, you'll evaluate three concrete steps that could be taken to reduce CO_2 emissions.

6.2 Background Information

6.2.1 The Problem

For decades, efforts to control CO_2 emissions have been hindered by climate deniers who have either refused to recognize that the climate is changing or focused the blame on a number of other factors, including variations in the Earth's orbit, changes in solar activity, or volcanic eruptions. Other climate myths, such as "it's part of a natural cycle" or "CO_2 is good for plants," have also been used to justify inaction.

Based on the body of work presented in the new discipline of attribution science, there is no longer serious debate among the scientific community: (a) climate change is occurring; (b) this change will continue to impact the biosphere; and (c) human activities are the primary driver of these changes (Fig. 6.3). While many have been raising the alarm, humankind so far has been unsuccessful at slowing the pace of change. As evidence of current and potential climate impacts continues to emerge, it is apparent that much more needs to be done now if the worst impacts are to be avoided.

6.2.2 The Role of Climate Change

While CO_2 absorbs less heat per molecule than some other GHGs like methane (CH_4) or nitrous oxide (N_2O), it is more abundant, stays in the atmosphere much longer, and is responsible for about two thirds of witnessed temperature increases. This, in combination with the fact that human activities have increased its concentration in the atmosphere significantly, makes CO_2 the most important GHG to control to tackle climate change.

There are a number of reasons why reducing carbon emissions globally has proven to be challenging:

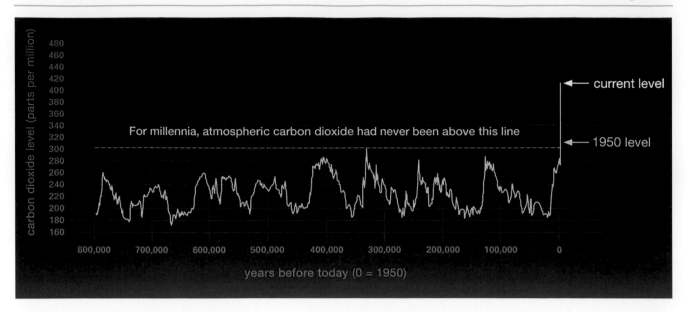

Fig. 6.1 Atmospheric samples contained in ancient ice cores, along with more recent direct measurements, provide evidence that atmospheric CO_2 has increased dramatically since the Industrial Revolution. (Source: Climate.NASA.gov {Public Domain})

Fig. 6.2 Historical CO_2 (right axis) and reconstructed temperature anomaly (Deviation from the 100-year mean) based on Antarctic ice cores. (Source: Leland McInnes (CC BY SA 3.0) via Wikimedia Commons)

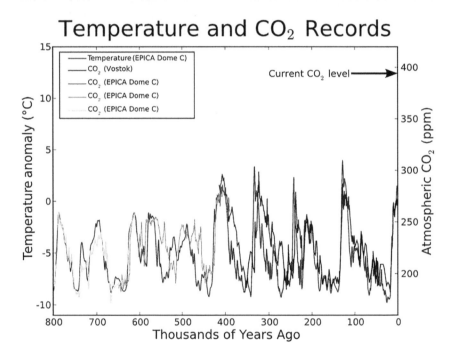

- As climate change is a global issue, efforts made by any one nation must be duplicated globally. This requires cooperation and willingness to make sacrifices across political and ideological boundaries. This is particularly important since many of the worst impacts of climate change affect countries that have contributed little to the problem.
- Sources of GHGs like CO_2 are literally everywhere, from the tailpipes of gas-powered vehicles to organic carbon losses from tilled agricultural fields to wildfires.

- The scale of potential disruption to ecological, societal, and economic systems is breathtaking. Ongoing and future impacts of climate change will touch virtually every aspect of the biosphere.
- Climate change has become politicized. Fossil fuel interests, among others, in the private sector stand to lose economically as nations move to greater reliance on renewable energy sources.
- Limiting carbon emissions requires us to change. It isn't a problem that requires action in a far-off place or steps

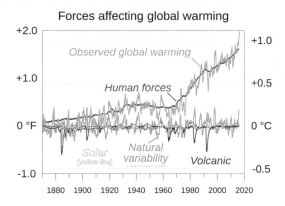

Fig. 6.3 Modeled forces affecting global temperature. (Source: NCA4 (2017) RCraig09 (CC BYSA 4.0) via Wikimedia Commons)

implemented only at the governmental level. It requires each of us to consider how we might reduce our own carbon footprint through lifestyle changes and intentional choices.

However, these are not insurmountable barriers. They just require all of us, environmental scientists, government leaders, and global citizens, to view the problem through a new lens.

6.2.3 Solutions

While the success of efforts to coordinate actions across nations has been limited, other entities, ranging from individual cities to the United Nations, have stepped up their own efforts to find and implement solutions. Approaches taken to address climate change fall into several categories:

Adopt Carbon Emissions Policies, Regulations, and Incentives Policies designed to reduce such emissions range from taxes on carbon released from the burning of fossil fuels to cap-and-trade policies that incentivize industries to limit carbon emissions. Homeowners can also be given incentives to upgrade their home's energy efficiency to reduce energy use and the accompanying release of GHGs.

Biologically Sequester Carbon Carbon sequestration is a natural process during which carbon is taken up from the atmosphere and incorporated into living organisms or natural components of the planet. Oceans, forests, and soils currently store significant carbon reserves (Fig. 6.4), which can also make them significant sources of CO_2 if disturbed. However, approaches such as large-scale reforestation and regenerative agriculture have the potential to sequester significant amounts of CO_2.

Carbon Capture and Storage Technologies New approaches for removing CO_2 from the atmosphere such as direct air capture (DAC) and geo-engineered molecules that can filter incoming solar radiation abound, but they are risky due to the complexity of the Earth system and uncertainty about how such measures might have cascading impacts.

While it is encouraging to see the wide range of initiatives being considered to tackle rising GHG emissions, the limited ability of governments to coordinate efforts leaves the current burden of action on communities, organizations, and individuals.

6.2.4 Unit Challenge

While reducing CO_2 emissions is a global challenge, many efforts occur at a smaller scale. This effort is being led by a large group of stakeholders representing various expertise, perspectives, sectors, and communities.

Similar to this effort, your Unit Challenge is to work with a diverse team **to identify steps to reduce your state's CO_2 emissions by 35% by 2030**. Your working group has specifically been asked to evaluate possible approaches to reduce net carbon flux statewide (e.g., either reducing CO_2 emissions or sequestering CO_2 from the atmosphere or both). Your group's challenge is to identify the approach that makes the most sense from environmental, social, and economic perspectives. You will present this as a one-page Fact Sheet to be included in the state's overall Climate Action Plan.

6.2.5 The Scenario

Your state has committed to an aggressive carbon-reduction program. Assume that it offers a number of opportunities and challenges when it comes to reducing CO_2 levels. It has a temperate climate with cold winters which require substantial insulation of buildings. Historically, your state was heavily forested, and ample land exists to support new forests. The citizens of the state are generally supportive of alternate energy sources like wind and solar, and polls suggest strong support for steps to address climate change.

After reviewing the various carbon-reduction strategies that might be appropriate for your state, your group decides to focus on comparing the potential of the following three:

1. **Residential energy efficiency.** An important strategy for fighting climate change is to reduce the amount of energy we consume in our daily lives. A prime example is the energy required to heat and cool our homes. More and better insulation of building cavity walls, solid walls, and

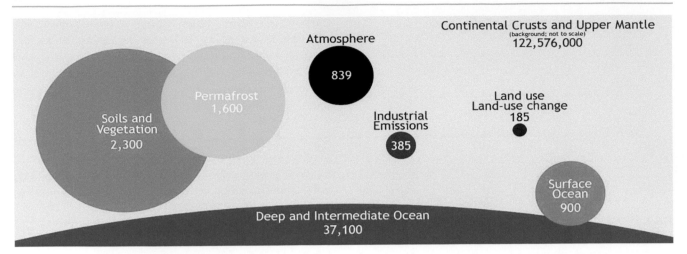

Fig. 6.4 Global carbon stocks (carbon stored in pools) shown in gigatons. (Source: USFS, Climate Change Resource Center {Public Domain})

lofts/attics would help reduce energy consumption and lower GHG emissions (Totland et al. 2019).

Levy et al. (2016) looked at the picture in the United States, focusing on both carbon reductions and human health benefits that would result from improved US residential energy efficiency. They estimated that increasing insulation to international code levels in all single-family homes in the United States would lead to reductions of 80 million tons of CO_2 released from power plants.

Specific recommendation: Increase residential insulation in all 50,000 single-family homes in the state built prior to 1950 to meet the International Energy Conservation Code 2012 levels. Because of the state's cold climate, insulation with an R value of 35 will be required.

2. **Solar farms (photovoltaic power stations).** A solar farm is a large solar array (Fig. 6.5) that generates renewable energy from the sun which is then routed to the power grid to offset the need for fossil fuel-sourced energy. Becoming more popular because landowners can generate additional income from unused land, industrial-scale solar farms have many proponents but also generate concern in local communities. In some cases, land may be purchased outright for solar farm installation, but more often, landowners lease their land to large corporations to install and manage the solar farm.

Specific recommendation: Install a solar farm on 500 ha of leased land.

3. **Reforestation.** Forests sequester more carbon than any other land use. In the United States alone, forests currently store 25 years' worth of CO_2 emissions. But due to tree mortality, wildfires, timber harvesting, and development, conversion of forests can also result in a net release of carbon. Trends in forest carbon sequestration/emission over the past few centuries show how rapidly we can lose

forest carbon, while at the same time, how sustainable management and reforestation efforts (Fig. 6.6) can restore some of those "lost" stocks. Globally, scientists estimate that the Earth could support another 900 million hectares (ha) of forests, which could reduce atmospheric carbon by about 25% (Bastin et al. 2019).

Specific recommendation: Increase forest cover statewide by 10,000 ha using a mix of state-owned properties and community partnerships on town and municipal lands.

You will discover that each of these options has various pros and cons and different costs and benefits. Your job is to evaluate these options and recommend an approach that will perform best over time under the conditions expected with climate change.

6.2.6 Relevant Facts and Assumptions

- The state has identified 1000 single-family homes needing R 35 insulation.
- The state has 500 ha available for solar farms.
- There are 10,000 ha of state/municipal-owned land that could be reforested.
- Currently, the state's electrical demand is met by gas-fired power plants that release 1.5 Mt of CO_2 per year.
- The average cost of insulating a home in your state to the desired level would be $3200 per home.
- 1 ha of solar panels costs about $4200 to install.
- Reforestation practices vary widely, depending on planting density, tree species. and tree age at planting. For our purposes, assume that reforestation efforts include a one-time cost of $1600/ha.

Fig. 6.5 Solar farms concentrate photovoltaic cells to feed renewable energy into the electric grid. (Source: junilly (CC BY 3.0) via Wikimedia Commons)

Fig. 6.6 A multicountry-led effort called the African Forest Landscape Restoration Initiative (AFR100) is working to reforest 100 million ha of land in Africa by 2030. (Source: Andrea Borgarello for TerrAfrica/World Bank, via NASA. gov {Public Domain})

6.2.7 Build Your Foundational Knowledge

Below are web sources that provide additional information about each of the solutions you're considering for this Unit Challenge. This information can build a critical foundation to help you evaluate each option and support your final choice. After reviewing each source, be prepared to answer questions in the Preparation Assessment

Quiz and to summarize any information relevant to your Unit Challenge.

Residential energy efficiency:
Halfway There: Energy Efficiency Can Cut Energy Use and Greenhouse Gas Emissions in Half by 2050.

Solar farms:
How solar panels reduce your carbon footprint

Reforestation:
Afforestation and Reforestation: Restoring trees to ecologically suitable landscapes

Final Product: A one-page Fact Sheet summarizing the proposed approach to help meet the state's carbon targets, as well as justifying its implementation as a part of the larger Climate Action Plan. Consider your audience, including state officials and the general public. Be sure to demonstrate how the proposed solution will stand up to the challenges of climate change.

6.2.8   Preparation Assessment Quiz

Are you ready to tackle your challenge? At this point you should understand the basic environmental principles and ecological processes involved in this environmental problem. Consider the following questions. If you are comfortable with answering these, then you are ready to head into Discovery, Analysis, and Solutions activities.

- List and "debunk" several common climate change myths.
- What are the primary barriers to reducing CO_2 emissions globally?
- What are some common policy approaches for reducing carbon emissions?
- How can natural resources be managed to sequester carbon?
- What are some promising technological approaches for sequestering and storing atmospheric carbon?
- According to the ACEEE web site, how great are the average energy savings for a home retrofit?
- OVO's article on solar energy notes that the medium installation cost of a 4 kW system has dropped by how much since 2010?
- According to the Economist article on afforestation and reforestation, restoring trees to the landscape could remove how much CO_2 annually?
- For each of your proposed solutions, are there any additional benefits that might arise from their implementation that might not be directly related to your Unit Challenge?

- For each of your proposed solutions, are there any negative unintended consequences that might result from their implementation?
- What additional information did you glean from your web sources that might help inform your Unit Challenge?

6.3   Controlling Carbon: Discovery

Specific Skills You'll Need to Review: Navigating the Scientific Literature, Science Communication, Problem-Solving

6.3.1   Independent Research

(Key Skill: Navigating the Scientific Literature)

To better understand the potential for each of your carbon-reduction options to help the state reach its stated goal of "35 by 30," you first need to examine the literature to see what others have found. Conduct a search of the peer-reviewed scientific literature focused on the solution you have been assigned, and identify one research paper that focuses on your assigned approach.

Prepare a summary of the article you selected that includes the following:

- **Citation**
- **Main topic**: Stick to a few words, likely pulled from the title.
- **General summary**: A few bulleted sentences summarizing the research question it addresses and approaches it takes.
- **Methods**: How did they approach their research question?
- **Location**: Where was the work done?
- **Conclusions**: Concise list of the findings, specifically capturing the take-home message.
- **Location**: Where was the work done?
- **Relevance**: How might this study help inform your Unit Challenge? Feel free to make a bulleted list of information you may want to include later.

6.3.2   Literature Share: Reciprocal Instruction

(Key Skill: Scientific Communication)

Share   In small groups, share and critique the research article you found. Keep in mind that your peers have not read this article, and it is your job to convey the key information to them. Note the items that will be important to consider when you are developing your final recommendations.

Critique Evaluate how these studies might help inform your Unit Challenge, considering the following:

- Source (Quality of the work or bias of the authors)
- Methods (Did their methods sufficiently address the research question?)
- Conclusions (Did the results justify the conclusions made?)
- Relevance (Can these findings be applied to your challenge?)

Based on your critiques, choose one article to share with the larger class, along with the key information that may be useful in deciding on a solution to propose.

6.3.3 Think-Compare-Share

(Key Skill: Problem-Solving)

Now that you have more information about possible solutions for this unit's challenge, you need to **develop a more formal problem definition** to guide your work throughout the rest of the exercises.

Think Start by working independently to develop a specific Problem Statement to guide the remainder of your work. Problem Statements provide the relevant information and boundaries to make the issue something you can effectively assess and tackle. The basics of a formal Problem Statement include the following:

Problem Statement: A short, concise statement summarizing the issue that includes the following:

- A **description** of the specific objective or outcome you hope to achieve (What is the actual problem?)
- **Justification** for addressing the problem (Why does this problem matter?)
- Potential **sources** or **causes** of this problem (What is the cause you need to address?)
- The **metrics** you will use to assess the status of the problem (How will you know if you are making a difference in the problem?)
- The **desired outcome** for these metrics (What is the end goal or ideal state?)
- Potential **solutions** to consider (How might you attempt to achieve this goal?)

Compare/Share Now return to your small group to share your Problem Statements. Use each of your ideas to develop a joint Problem Statement that contains all key information and is concise, clear, and well written.

Unit Discovery Summary *Submit a final Problem Statement that succinctly captures the key information to guide your work on this Unit Challenge.*

6.3.4 Reflecting on Your Work

(Key Skill: Personal Reflection)

After your work in Discovery, you should have a better idea of the problems you face and have produced a Problem Statement you can use to tackle the Unit Challenge. Take a moment to reflect on this work. Consider the following prompts but feel free to expand on any to best capture your learning experience and better inform your next steps.

- Of the skills you practiced in Discovery, which were the most challenging? Which were the most interesting?
- How were you most comfortable working during these exercises? In small groups, independently, or with the larger class? Why? How does your choice reflect your personality type and leadership style?
- Was your Problem Statement strictly focused on the environmental problem of controlling carbon emissions, or did it also consider important social and economic considerations? How might a singular focus on the environmental aspects limit your ability to identify truly sustainable solutions?
- You've been given three viable solutions to assess as a part of this case study. But this is not an exhaustive list of options or even necessarily the best possible course of action for every scenario. Take a moment to "think outside the box." Are there any other possible solutions you think would be worth exploring? Describe one that you think would be worth pursuing.

6.4 Controlling Carbon: Analysis

Specific Skills You'll Need to Review: Quantitative Literacy, Sustainability Science

Review your Background and Discovery sections before beginning the Rotating Station exercises below. While you focused on one potential solution in your Independent

Research in Discovery, keep an open mind as your work through Analysis activities.

6.4.1 Rotating Stations

(Key Skill: Quantitative Literacy)

At each of the following stations, you will review data that are relevant to the three potential solutions you're considering. Spend some time working through the analyses at each station to learn more about this issue and possible solutions for your Unit Challenge.

Be sure to write down one finding at each station that will help inform your selection of a solution.

Station 1: Residential Energy Efficiency Investing in residential insulation can result in substantial energy savings. Consider Fig. 6.7 from Levy et al. (2016) who estimated carbon reduction from residential efficiency measures implemented across the United States. This was focused on energy use for home heating. They found dramatic differences in CO_2 emission reductions across states and energy sectors.

Based on the information displayed in Fig. 6.7, answer the following:

- Both the total amount (e.g., gallons, etc.) and percent of savings (hashed legend) are shown for each state. Are the two always correlated across all states (e.g., darkest blue for amount also has the highest percentage savings)? Why or why not?
- Energy savings associated with these home heating/cooling efficiency efforts differ by energy source (e.g., electricity, natural gas, fuel oil). What might account for these differences?
- Where have states achieved the greatest overall percent reduction in CO_2 emissions (bottom panel) as a result of efficiency measures? Which states have the least percent emissions reduction? What does this tell you about where these types of residential efficiency measures may be most effective?
- Consider the state your challenge is focused on. What percent savings in overall residential energy use might you expect to achieve with insulation efforts?

Record any relevant Station 1 findings

Station 2: Solar Farms In order to estimate the potential for solar panels to mitigate energy-related carbon emissions, the International Renewable Energy Agency (IRENA) (2019) analyzed how 2018 levels of energy-related CO_2 emissions could be reduced by integrating various renewable

energy sources by the year 2050. The overall goal is to reduce carbon-based energy-related CO_2 emissions to 9.8 Gt CO_2/year (green in the figure below).

Answer the following about Fig. 6.8:

- Solar PV could contribute 4.9 Gt of CO_2 emissions reductions in 2050, representing 21% of the overall emission reductions needed to meet the Paris Accord climate goals. What proportion of the baseline 2018 total energy-related emissions would this replace? How does this proportion compare to other renewable energy sources?
- Some of the solar PV emissions savings would come from installations on homes. Homeowners will need to know how quickly investments in panels would pay off in electric bill savings. How long would it take to recover the capital outlay for installation of a 6-kW solar array on a residential building ($1000/kW for installation) that provides an average of 90 kWh/month (with an associated savings of 2.8 cent/kWh)?
- Now consider how the availability of suitable solar siting can influence your expected emissions reductions. Assume that a properly sited 1-ha solar farm can generate about 140 mWh annually, preventing the emission of about 60 tCO_2/year. However, this can be 70% lower at locations where local topography limits sun exposure. If you have the option to install a solar farm on 15 ha with ideal solar siting, or 20 ha of a more limited site (with the 70% reduction in energy generation), which would you choose?

Record any relevant Station 2 findings

Station 3: Reforestation Bernal et al. (2018) explored potential CO_2 removal by naturally regenerated forests, agroforestry, and mangrove restoration practices in various regions across the globe. Estimates were given for both short-term (0–20 years) and long-term (20–60 years) carbon sequestration potential for each of these approaches, with estimates for natural regeneration shown in Fig. 6.9.

Consider the figure above and answer the following:

- How do rates of carbon sequestration differ between years 0–20 (in green) and years 20–60 (in orange)? Why might the overall rate of sequestration differ between young and more mature vegetation?
- How do rates of carbon sequestration differ between aboveground and belowground carbon pools?
- Now consider how carbon sequestration varies in years 0–20 and in years 20–60 across regions. Where is overall sequestration highest? Where is it lowest? What might explain these patterns?

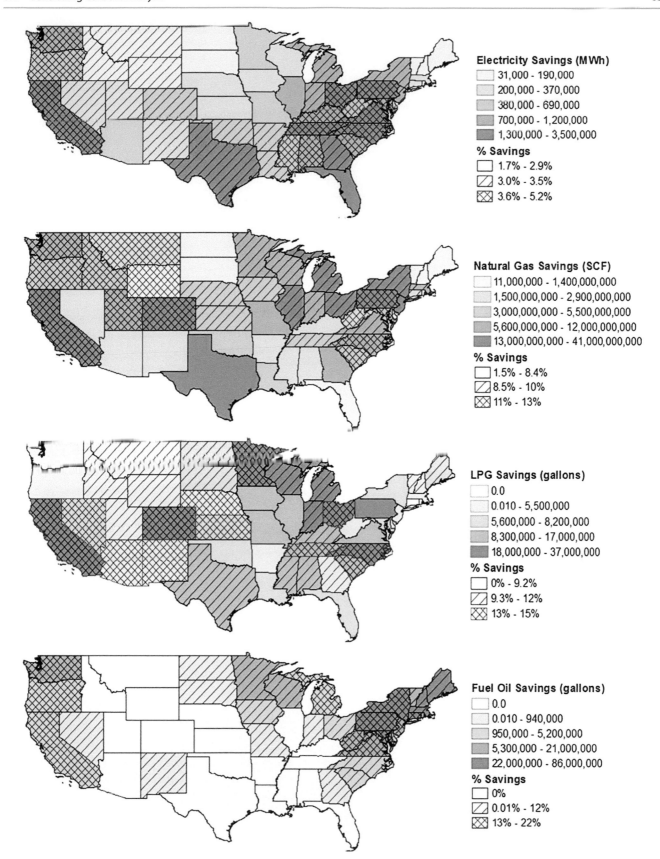

Fig. 6.7 Annual total and percent reduction in residential electricity and fuel consumption and overall CO_2 emissions reductions associated with increased insulation. (Source: Levy et al. 2016 (Open Access))

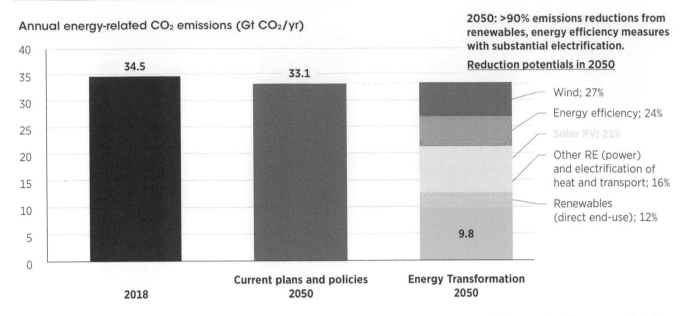

Fig. 6.8 Annual energy-related CO_2 emissions (Gt CO_2/year) under baseline (2018) and potential (2050) scenarios. (Source: modified from IRENA 2019 {Public Domain})

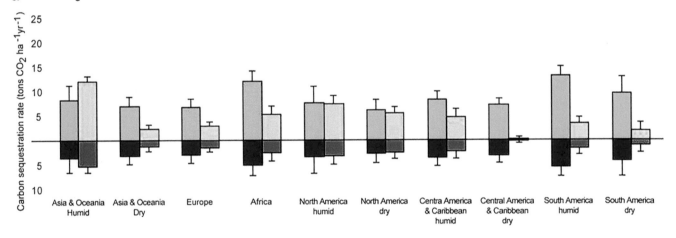

Fig. 6.9 Carbon sequestration rate of natural regeneration. Green colors represent rates during the first 20 years of tree growth (aboveground biomass in light green, belowground biomass in dark green), while orange colors represent rates during 20–60 years of tree growth (aboveground biomass in light orange, belowground biomass in dark orange). (Source: Bernal et al. 2018 (Open Access))

Record any relevant Station 3 findings

Station 4 To support your recommendations, you will have to consider both the effectiveness of various approaches to reduce carbon emissions and the relative cost of implementing those approaches. Consider the assumptions of relative costs and mean emissions reductions for your various climate action strategies shown in Table 6.1.

- Which Climate Action has the greatest overall net carbon reduction potential per ha?
- Which is the cheapest to implement per ha?

- What is the net cost per tCO_2 equivalent/ha for each climate action strategy in year 1 (e.g., 1 year of net emissions reductions)? Which strategy is the most cost effective based on these calculations?
- The cost is a one-time implementation expense, but the net reduction in emissions continues year after year at the same rate. Repeat your calculations for the net cost per tCO_2 equivalent/ha for each climate action strategy considering 10 years of emissions reductions. Repeat these again for a 25-year span on emissions reductions for each strategy. How do the differences in the per tCO_2 equivalent/ha reduction in emissions change among these three strategies?

Table 6.1 Estimates of the costs and effectiveness of various climate action strategies proposed in the Unit Challenge

Climate action	Cost per ha	Net emissions reduction (tCO$_2$ equivalent/ha/year)
Reforestation	1600	0.46
Solar Farm	4200	0.82
Residential Insulation*	3200	0.39*

*Assumes a single home per hectare

Record any relevant Station 4 findings

6.4.2 System Mapping

(Key Skill: Sustainability Science)

Your assigned solution may have a variety of direct and indirect economic, social, and ecological impacts that should also be considered. For example, upgrading residential insulation may require marketing to encourage homeowners' participation; reforestation may provide additional ecosystem services to the surrounding community; and the purchase of PV solar arrays may be opposed by local communities interested in maintaining their pastoral character.

Working with other class members assigned the same solution, develop a simple sustainability map that shows the various system connections across each of the three sustainability domains (ecological, economic, and societal). When you find connections between impacts that cross domains, draw a line to identify the connection (e.g., increased wildlife habitat in reforested areas (ecological) leading to increased hunting opportunities (social/economic)).

The goal is to think broadly about this larger system and how implementing actions in one domain (e.g., ecological) may impact components in another domain (e.g., economic or societal).

Use the following template (Fig. 6.10) to get started:

When each group has completed their basic sustainability map, come together as a class to compare system maps for the three possible solutions. This information will help inform your decision support work later in the unit.

Unit Analysis Summary *Based on your explorations, what have you learned that can help inform your choice of a solution? Do the data support the adoption of one of these potential solutions?*

6.4.3 Reflecting on Your Work

(Key Skill: Personal Reflection)

During your work in Analysis, you explored some of the research into possible solutions to help inform your decision. Take a moment to reflect on this work. Consider the following prompts but feel free to expand on any to best capture your learning experience and better inform your next steps.

- How did you feel working with data? Do you consider quantitative literacy a strength or an area for improvement for you?
- How important should science be in informing decisions around Climate Actions? Do you feel the data you examined support and justify actions to mitigate carbon emissions?
- Any solution should be examined using a sustainability science lens. Your goal is to reduce atmospheric CO$_2$ levels in your state. But any solution you implement will have other direct and indirect impacts. What are other possible economic, social, or ecological impacts (e.g., might employment be stimulated by a residential insulation program? Or would placement of solar arrays run into local opposition from residents?)? How should such considerations influence your decision?

6.5 Controlling Carbon: Solutions

Specific Skills You'll Need to Review: Problem-Solving, Decision Support, Communicating Science

Review your Unit Challenge and the major findings from Discovery and Analysis, including the sustainability map you created to highlight connections among possible solutions and the larger socioecological system.

6.5.1 Small Group Guided Worksheets

(Key Skill: Decision Support)

Decision support matrices can help break down the desired outcomes to reflect multiple criteria for consideration, and they allow you to compare how each possible solution achieves those desired outcomes. This not only helps inform decision-making, it also provides transparency in the decision process and justification to help you advocate for its adoption.

For the three possible solutions, you will evaluate how well each can achieve the following desired outcomes:

- Reduces net atmospheric carbon emissions
- Provides affordable outcomes

Fig. 6.10 Template sustainability map to help connect ecological, societal, and economic considerations associated with your potential solutions

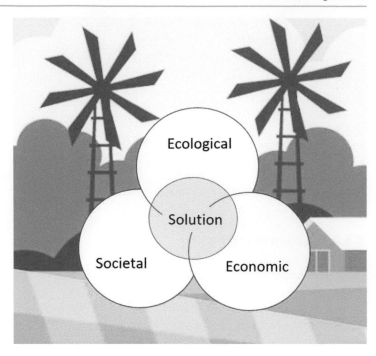

- Provides a lasting impact (not a short-term fix), even with a changing climate
- Is accepted by stakeholder groups
- Minimizes negative secondary impacts (e.g., on surrounding ecosystems or communities)

Considering your three potential solutions, develop a formal decision support matrix to compare and evaluate each approach using the template matrix below (Table 6.2). Getting the most out of the decision matrix requires a depth of knowledge about each of the three possible solutions. Below we list additional sources about each option. Please review each of these, paying particular attention to the two alternatives to your assigned approach.

Note that your group may have uncovered a novel solution not included in this list of three. You may choose to work through the structured decision matrix with your self-identified solution as a fourth solution option.

6.5.2   Additional Sources

1. Residential energy efficiency: Energy efficiency and the fight against climate change |World Economic Forum (weforum.org)
2. Solar Farms: The Potential of Solar Energy to Mitigate Climate Change – Off Grid World
3. Reforestation: In best-case reforestation scenario, trees could remove most of the carbon humans have added to the atmosphere | NOVA | PBS

Score each of your three solutions for each of the desired outcomes using a simple relative scale of 1 for least benefit to 5 for greatest benefit. Using a relative scale means you don't need to know exactly how well each solution meets the goal of each desired outcome, but you can use your judgment to assess how well each solution works compared to the others.

While this is a relative (subjective) scale, note that you will need to justify your scoring of each solution for each desired outcome.

Once each cell in your decision matrix has a relative score, calculate an average score for each solution. Based on this analysis, which is the "best" solution considering all your desired outcomes?

6.5.3   Role Playing

(Key Skill: Communicating Science)

In tackling environmental issues, you'll often find yourself working with groups of people with different perspectives and opinions about implementing a solution. You should use clear, concise communication to summarize your solution and justify its selection. To be effective, this must address concerns likely to be presented by various stakeholder groups and how the risks of taking these actions outweigh the risks of taking no action.

Your group will present and justify your chosen solution to the class with a particular emphasis on how it might benefit or impact one of the key stakeholder groups listed below. Be sure to include arguments that support this solution from the

Table 6.2 Basic decision matrix to compare the three possible solutions for your Unit Challenge. Include your justification for each solution's ranking for each desired outcome

	Desired Outcomes				
Solution Options	**Maximizes effectiveness (net emission reduction)**	**Minimizes overall cost to implement**	**Maximizes duration / longevity of effect**	**Maximizes stakeholder buy-in**	**Minimizes secondary impacts**
Insulation (energy efficiency)					
Solar farms (energy generation)					
Reforestation (sequestration)					
Justification for ranking					

Use a scale of 1-5 to score each option under each desired outcome category where 1 = does not achieve the desired outcome and 5 = completely achieves the desired outcome.

ecological, societal, and economic domains of sustainability science. Listeners will also be **assigned a stakeholder identity**, with an opportunity to **ask follow-up questions** after your presentation that reflect their unique concerns and perspectives. Your job will be to listen carefully and tailor your answers to this audience of stakeholders.

Key stakeholder groups include the following:

- Local residents who live near or own the properties where PV solar will be developed, where energy efficiency incentives will be focused, or where reforestation will occur
- Politicians and decision-makers concerned about potential public reactions
- State regulators responsible for preserving the state's environmental quality
- NGOs focused on reducing the impacts of climate change on the state
- Business interests from the energy sector

6.5.4 Reflecting on Your Work

(Key Skill: Personal Reflection)

During your work in Solutions, you explored several Climate Actions that could be taken to reduce CO_2 levels in your state. Take a little more time now to reflect on your find-

ings and the skills you practiced. Consider the following prompts but feel free to expand on any of them to best capture your learning experience and feelings about this issue.

- Reflect on your work today through a sustainability science lens. Does your solution address ecological, economic, and societal considerations? Which considerations do you think should carry the most weight in the decision? Why?
- How did you weigh solutions that might have the greatest environmental impact against those that are most likely to be implemented and maintained for long-term impact?
- How can environmental scientists work to show the value of healthy ecosystems and justify the costs of mitigation strategies?
- Science communication can be challenging, especially when working with diverse audiences. We need to craft our communication to match the interests and values of the target audience, but how do you do this when your audience contains a mix of stakeholder groups? How can you maximize the impact of your message to a diverse audience?

Unit Solution Summary Summarize and justify your final solution choice and outline how it addresses the direct challenge while also considering social, economic, and ecological impacts. Demonstrate that it will continue to meet the challenges posed by climate change.

6.6 Controlling Carbon: Final Challenge

As a part of this Unit Challenge, you were asked to write a one-page Fact Sheet summarizing your choice for effective Climate Actions to minimize net carbon levels by 2030. Your Fact Sheet should include the following components:

- Brief problem statement
- Recommended Climate Action with sufficient details to summarize the general approach
- Justification of this recommendation (e.g., long-term effectiveness given anticipated climate changes, implementation costs, other benefits provided, etc.). Be sure to use a sustainability lens to include considerations of direct and indirect ecological, social, and economic considerations
- Any obstacles the group might face trying to implement your solution

Consider using figures, graphics, and tables to help summarize the system and show how this solution is well suited to meet all desired outcomes.

Final Unit Challenge Submit your final recommendations in a one-page Fact Sheet using clear science communication designed for a lay audience.

References

Bastin JF, Finegold Y, Garcia C, Mollicone D, Rezende M, Routh D, Zohner CM, Crowther TW (2019) The global tree restoration potential. Science 365(6448):76–79

Bernal B, Murray LT, Pearson TR (2018) Global carbon dioxide removal rates from forest landscape restoration activities. Carbon Balance Manag 13(1):1–13

International Renewable Energy Agency (IRENA) (2019) Future of Solar Photovoltaic: Deployment, investment, technology, grid integration and socio-economic aspects (A Global Energy Transformation paper). Abu Dhabi.

Levy JI, Woo MK, Penn SL, Omary M, Tambouret Y, Kim CS, Arunachalam S (2016) Carbon reductions and health co-benefits from US residential energy efficiency measures. Environ Res Lett 11:034017

Totland M, Kvande T, Andre R (2019) The effect of insulation thickness on lifetime CO_2 emissions. IOP Conf Ser Earth and Environ Sci 323(012033):1–9

Core Knowledge
Human populations, Environmental justice, Coastal systems, Agricultural systems

7.1 Environmental Issue

Since 1900, global mean sea level has risen more than 20 cm (Rahmstorf, 2021). During this period, the rate of rise has accelerated and, from 2006 to 2018, sea level rose at 3.7 mm/year. Projections indicate that the rate of sea level rise will continue to accelerate, with estimates ranging from an additional 1–3 m over the next century. About 200 million people currently live in coastal areas projected to be below sea level over this time period. Even more live in vulnerable coastal locations where frequent and severe tidal flooding will disrupt livelihoods and basic subsistence.

Residents of densely populated developing nations with substantial coastlines are particularly threatened by the ravages of climate change, with a combination of climate-driven impacts likely to create an ever-increasing number of environmental refugees. Among those at greatest risk will be the millions living along the coastlines and rivers of Bangladesh (Fig. 7.1). With increased coastal and inland flooding, extreme climate events, and ongoing sea level rise, strategies to mitigate the risks and help Bangladeshi communities adapt are essential to supporting vulnerable populations.

In this unit, you'll assess three approaches that may help coastal residents cope with the onslaught of climate-driven sea level rise along Bangladesh's shorelines.

7.2 Background Information

7.2.1 The Problem

Sea Level Rise One of the starkest effects of climate change is the anticipated rise in sea level worldwide. This occurs for two main reasons—the expansion of the ocean as it warms and the increased melt from ice sheets, ice caps, and glaciers. Along with alarming threats to coastal communities, infrastructure, food production, economies, and ecosystems, this rise has implications for available freshwater, as rising sea levels drive saltwater into freshwater aquifers.

Global mean sea level has risen about 20 cm since 1900, with about one third of that coming in just the last two and a half decades (Fig. 7.2). In 2019, global mean sea level was 8.6 cm above the 1993 average—the highest annual average in the satellite record (1993 present). In just 1 year, from 2018 to 2019, global sea level rose 0.6 cm. New concerns about the stability of Antarctica's Thwaites or Doomsday Glacier were raised late in the summer of 2022, with the possibility of as much as a meter of sea level rise occurring if the underwater base of the glacier disintegrates.

While increases in sea level vary across the globe due to ocean floor topography and prevailing wind patterns, when sea levels rise as rapidly as they have been, even a small increase can have devastating effects on coastal habitats and human populations farther inland.

Climate Refugees Displacement of humans by climate change is already occurring on a global scale. From the flooding of islands like Vanuatu in the South Pacific to the migrations caused by food insecurity from crop losses in Central America, climate-related changes are already creating large numbers of environmental refugees.

Marginalized populations in developing nations are often most at risk. A report by the World Bank Group (Rigaud et al. 2018) estimated that by 2050, more than 140 million residents of sub-Saharan Africa, South Asia, and Latin America will be on the move because of climate-related disruptions. Given inadequate support systems in many nations, such mass movement will likely pose a grave threat to global stability.

© The Author(s), under exclusive license to Springer Nature Switzerland AG 2024
J. Pontius, A. McIntosh, *Environmental Problem Solving in an Age of Climate Change*, Springer Textbooks in Earth Sciences, Geography and Environment, https://doi.org/10.1007/978-3-031-48762-0_7

Fig. 7.1 Houses are nearly submerged due to flooding in Sirajganj, Bangladesh. (Credit: Moniruzzaman Sazal/ Climate Visuals Countdown, (CC BY SA 2.0) via Flickr)

Fig. 7.2 Seasonal (3-month) sea level estimates from Church and White (2011) (light blue line) and University of Hawaii Fast Delivery sea level data (dark blue). The values are shown as change in sea level in millimeters compared to the 1993–2008 average. (Source: NOAA Climate.gov {Public Domain})

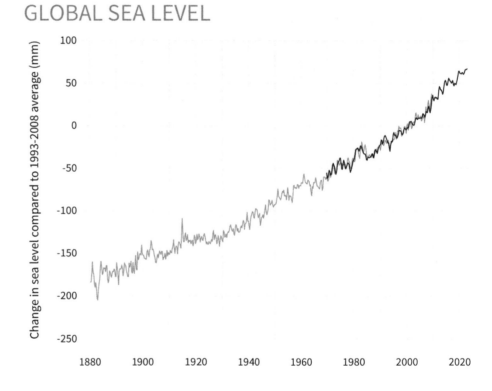

GLOBAL SEA LEVEL

For some, this movement involves migrating across national borders under the guidelines and supports afforded to refugees. However, most will be unable to afford passports or travel and will instead be displaced to nearby locations within their own borders bringing only what they can carry. Rather than a sudden migration due to catastrophic disaster, the displacement of climate-impacted populations can be more insidious. Even for those who are not formally displaced to a new location, climate change can worsen living conditions, damage natural resources such as drinking water, and reduce the ability to raise crops and livestock (Fig. 7.3).

In such conditions, climate change can act as a threat multiplier, exacerbating existing tensions and increasing the risk of conflicts.

Bangladesh at Risk Bangladesh occupies the largest delta in the world, with almost one third of its land less than 5 m above sea level. With its high population density and vulnerable infrastructure, Bangladesh is exceptionally at risk from climate change-related threats like sea level rise (Fig. 7.4). The nation presents a stark reminder of the daunting challenge facing global society. According to a report by the

Fig. 7.3 NOAA's interactive sea level rise viewer shows land area along the US Gulf of Mexico's coastline that will be under water (in light blue) with a half meter increase in sea level. (Source: NOAA {Public Domain})

Environmental Justice Foundation (EJF) (2018), as many as 18 million Bangladeshi will have to move by 2050 because of sea level rise alone. Many of the displaced will flee to the slums of Dhaka, the nation's capital. Much of the landscape in and around Dhaka will also be at risk from coastal flooding in future decades, likely creating even more climate refugees over time with fewer options for relocation.

A stark example of the extent of the problem is the status of Bangladeshi living on chars, the nation's river islands. As storm flows and river flooding increase with climate change, the 4 million Bangladeshi living on chars are especially at risk, as these islands can be completely eroded in a matter of days or weeks (EJF, 2018). Even cities like Chittagong, the country's largest port, are at risk. The city's 4 million inhabitants routinely face coastal flooding that is expected to increase with sea level rise. Estimates are that 27 million people could be permanently displaced from their homes in Bangladesh by 2100.

The combination of a highly vulnerable population and a host of climate-related threats makes Bangladesh an area of great concern in the coming decades. The statistics paint a grim picture. Consider the following statements and projections included in the EJF report:

- By 2050, one in every seven Bangladeshi will be displaced by climate change.
- Coastal water supplies serving 33 million people have been contaminated by salt.
- 28% of Bangladesh's population lives along its coast.
- Projected sea level rise by 2050 will result in a loss of 11% of the nation's land.

7.2.2 The Role of Climate Change

While there are clear links between increased global temperatures and sea level rise, the direct drivers of this change are complex. About a third of the increases witnessed so far can be attributed to thermal expansion from the warming ocean. Melting glaciers and ice sheets, as well as changes in groundwater storage, also contribute significant amounts of freshwater to the oceans.

However, sea level rise is not only the climate stressor on human populations in these vulnerable coastal areas. In nations like Bangladesh, other climate-related threats compound the problem. More and stronger cyclones and inland riverbank erosion during monsoons will contribute to the loss of infrastructure, while saltwater intrusion of water sup-

Sea Level Risks - Bangladesh

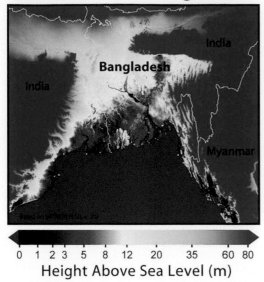

Height Above Sea Level (m)

Fig. 7.4 A topographic map emphasizes portions of Bangladesh that are near sea level and hence could potentially be vulnerable to sea level rise. As large portions of Bangladesh are near sea level, sea level rise here has the potential to displace tens of millions. (Source: Robert A. Rohde (CC BY SA 3.0) via Wikimedia Commons)

plies and loss of agriculturally productive lands due to flooding and salinization will add to the stress placed on coastal residents (Fig. 7.5). Cyclones and storm surge will impact marine fishing, infrastructure, human life, and property, while diseases like malaria and dysentery will likely become more widespread in a warmer, wetter climate. With one third of Bangladesh's population at risk of displacement from threats like sea level rise and coastal erosion, it's not surprising that the nation has been ranked among the top 10 nations in the world most affected by extreme weather events in recent years.

7.2.3 Solutions

Many governments and organizations have been working on identifying ways to help vulnerable coastal communities adapt to rising sea levels. At some locations, this includes building sea walls to keep the rising water out. In others, infrastructure improvements like raising roads can help avoid impacts. But for many low-lying areas like Bangladesh, where agriculture is also concentrated along the coastline, increased flooding and saltwater intrusion require more comprehensive strategies to help communities adapt.

For example, to offset the damage being done to agricultural productivity, alternatives like saltwater-tolerant crops like rice and conversion to aquaculture are being implemented. Farmers are using floating seedbeds, allowing crops

to rise and fall with floodwaters (Fig. 7.6). To minimize the risk to housing, flood-resistant and flood-proof housing is being constructed. To combat increasing salinity in water supplies, rainwater harvesting and other freshwater storage options are being explored. To protect valuable farmland, embankments and tree plantings can act as natural barriers to sea level rise.

Bangladesh faces severe challenges from climate change. Not only are there many adverse impacts occurring simultaneously, with many most likely to get worse over time, but the cause of the problems is a global one. Bangladesh has not created this problem and cannot significantly lower global greenhouse gas (GHG) levels to solve it. To avoid displacing millions from their homes, the best option is to help these communities adapt and be resilient in the face of the changes occurring.

7.2.4 Unit Challenge

As a member of the UN's Human Rights and Climate Change group, you have been assigned to work with a small agricultural community lying along Bangladesh's southeast coastline. This community has been specifically threatened by tidal flooding and increasing salinization of their soils. Recent decreases in the productivity of their traditional rice crops suggest that, without help, the community may have to relocate. Your Unit Challenge is **to identify ways this community can adapt their agricultural practices to be resilient to rising sea levels.**

7.2.5 The Scenario

Your group has identified a target community to work with. It is small enough to be manageable, but your group believes that the solutions you come up by working with the community may be more broadly applied along Bangladesh's threatened coastline.

A key throughout your effort will be to find a solution that will have a high likelihood of success. After reviewing the fundamentals of the small community (e.g., location, elevation above sea level, conditions of the soils and rice crops), you identify three options that seem to offer the best chance of success:

1. **Salt-tolerant rice.** Rice is the world's second most grown staple crop after wheat, but its production is adversely impacted by increased salinity levels (Fig. 7.7). While rice has some ability to adapt to saline conditions, development of salt-tolerant genotypes has been underway for years. Recently, a variety of salt-tolerant strains of rice

Fig. 7.5 Polders help hold back and direct water but also support shrimp and rice farming in the Khulna District, Bangladesh. (Photo: Al Helal/REACH (CC BY 2.0), via Flickr)

Fig. 7.6 Floating gardens allow cultivation in areas that are often flooded. (Source: Nazmulhuqrussell (CC BY 3.0) via Wikimedia Commons)

Fig. 7.7 Rice paddies in Bangladesh are at risk of saltwater intrusion with rising sea level. (Photo: David Stanley, (CC BY 2.0) via Flickr)

have been developed for use in coastal areas of Bangladesh. Experimental work continues to determine the best strain for local conditions.

Specific recommendation: Work with the community to identify stable sources of salt-tolerant rice strains.

2. **Salt-based aquaculture**. Local farmers could convert their current rice-based agricultural operation to salt-based aquaculture (black tiger shrimp). Saltwater intrusion along Bangladesh's coastline has presented farmers an opportunity to raise saltwater species such as shrimp using aquaculture (Fig. 7.8). In addition to reducing poverty and enhancing food security, aquaculture can provide a number of economic benefits. However, a host of socioeconomic and environmental issues around shrimp farming persist.

Specific recommendation: Replace existing rice crops with an aquaculture operation based on the black tiger shrimp.

3. **Polder construction**. Polders are low-lying lands that have been protected from the sea by dikes. This approach to land protection was perfected in the Netherlands, but there have been more than 100 polders in use in Bangladesh since the 1960s and 1970s, protecting roughly half of Bangladesh's coastal area. The success of polder systems is dependent on careful construction, operation of sluice gates, and maintenance of internal canals.

Specific recommendation: Construct a series of dikes surrounding the community to protect existing rice crops from sea level rise.

You will discover that each of these options has various pros and cons and different costs and benefits. Your job is to evaluate these options and recommend an approach that will perform best over time under the conditions expected with climate change.

7.2.6   Relevant Facts and Assumptions

- The community in question lies on 8.6 ha and has a population of 100, with 10 farms that cover 4 ha.
- All farmers have agreed to pool their resources to support the one best alternative approach.
- Black tiger shrimp (*Penaeus monodon*) is the most popular choice for saltwater farming. Capital costs to set up the system are $15,000 USD with additional annual maintenance costs of $250 US/ha. About 50 kg of shrimp can be produced/ha/yr with average yearly production costs of about US $3.50 per kg. Farmers can expect to earn $14/kg.
- Assume that annual salt-tolerant rice seed expenses (including seed, nursery preparations, and seedling development) average US $58/ha/year and annual labor for maintaining and harvesting costs an additional $200/ha. Annual earnings of $500/year/ha can be expected from existing fields with no additional modifications.
- Polders vary widely in size and cost. Assume that a polder system designed solely to protect the 4 ha of agricultural fields would cost about US $ 10,000 to build and maintain. Assume no additional costs to farmers to grow traditional rice crops once polders are constructed and annual earnings of $650/ha/year.
- The Bangladesh government and the United Nations will provide $25,000 to cover the startup costs of the project. After that, revenue generated must cover operational costs.

Fig. 7.8 Freshwater prawns cultured in ponds help communities earn additional income along with other fishes in *the* same pond. (Photo: Winrock International. (CC BY 2.0) via Flickr)

7.2.7 Build Your Foundational Knowledge

Below are web sources that provide additional information about each of the solutions you're considering for this Unit Challenge. This information provides a critical foundation to help you evaluate each option and support your final choice. After reviewing each source, be prepared to answer questions in the Preparation Assessment Quiz and to summarize any information relevant to your Unit Challenge.

Salt-tolerant rice:

Advances and Challenges in the Breeding of Salt-Tolerant Rice

Saltwater aquaculture:

How Bangladesh's Shrimp Industry is Driving a Fresh Water Crisis

Polder construction:

The polder promise: Unleashing the productive potential in southern Bangladesh

Final Product: A one-page Fact Sheet summarizing the issue, detailing your solution, and justifying your choice. Consider the need to incorporate the perspectives and interests of the local community and specifically the farmers you are working with. Be sure to demonstrate how your proposed solution will stand up to the challenges posed by climate change to provide lasting solutions for this community.

7.2.0 Preparation Assessment Quiz

Are you ready to tackle your challenge? At this point you should understand the basic environmental principles and ecological processes involved in this environmental problem. Consider the following questions. If you are comfortable with answering these, then you are ready to head into our Discovery, Analysis, and Solutions activities.

- What are the primary drivers of sea level rise?
- What is an environmental migrant?
- Why is Bangladesh particularly vulnerable to the effects of climate change?
- What are some ways communities are working to adapt to climate change in Bangladesh?
- What do Qin et al. suggest as a solution to the challenge of growing rice in salty waters?
- According to Siddique, how much did income from Bangladesh's shrimp exports rise between 1973 and 2012?
- According to the "Polder Promise," how many polders exist along Bangladesh's coastal zone and what percentage of the zone is poldered?
- For each of your proposed solutions, are there any additional benefits that might arise from their implementation that might not be directly related to your Unit Challenge?
- For each of your proposed solutions, are there any negative unintended consequences that might result from their implementation?

- What additional information did you glean from your web sources that might help inform your Unit Challenge?

7.3   Climate Migrants: Discovery

Specific Skills You'll Need to Review: Navigating the Scientific Literature, Science Communication, Problem-Solving

7.3.1   Independent Research

(Key Skill: Navigating the Scientific Literature)

To better understand the various approaches proposed for supporting at-risk coastal agricultural communities in Bangladesh, you first need to examine the literature to see what others have discovered. Conduct a search of the peer-reviewed scientific literature focused on the solution you have been assigned and identify one research paper that focuses on your assigned approach.

Prepare a summary of the article you selected that includes the following:

- **Citation**
- **Main topic**: Stick to a few words, likely pulled from the title.
- **General summary**: A few bulleted sentences summarizing the research question it addresses and approaches it takes.
- **Methods:** How did they approach their research question?
- **Location**: Where was the work done?
- **Conclusions**: Concise list of the findings, specifically capturing the take-home message.
- **Relevance**: How might this study help inform your Unit Challenge? Feel free to make a bulleted list of information you may want to include later.

7.3.2   Literature Share: Reciprocal Instruction

(Key Skill: Scientific Communication)

Share In small groups, share and critique the research article you found. Keep in mind that your peers have not read this article, and it is your job to convey the key information to them. Note the items that will be important to consider when you are developing your solution to the challenge.

Critique Evaluate how these studies might help inform your Unit Challenge. Consider the following:

- Source (Quality of the work or bias of the authors)
- Methods (Did their methods sufficiently address the research question?)
- Conclusions (Did the results justify the conclusions made?)
- Relevance (Can these findings be applied to your challenge?)

Based on your critiques, choose one article to share with the larger class, along with the key information that may be useful in deciding on a solution to propose.

7.3.3   Think-Compare-Share

(Key Skill: Problem-Solving)

Now that you have more information about possible solutions for this unit's challenge, you need to **develop a more formal problem definition** to guide your work throughout the rest of the exercises.

Think Start by working independently to develop a specific Problem Statement to guide the remainder of your work. Problem Statements provide the relevant information and boundaries to make the issue something you can effectively assess and tackle. The basics of a formal Problem Statement include the following:

Problem Statement: A short, concise statement summarizing the issue that includes the following:

- A **description** of the undesired condition or change that you hope to achieve (What is the actual problem?)
- **Justification** for addressing the problem (Why does this problem matter?)
- Potential **sources** or **causes** of this problem (What is the cause we need to address?)
- The **metrics** you will use to assess the status of the problem (How will you know if you are making a difference in the problem?)
- The **desired outcome** for these metrics (What is the end goal or ideal state?)
- Potential **solutions** to consider (How might you attempt to achieve this goal?)

Compare/Share Now return to your small group to share your Problem Statements. Use each of your ideas to develop a joint Problem Statement that contains all key information and is concise, clear, and well written.

Unit Discovery Summary *Submit a final Problem Statement that succinctly captures the key information to guide your work on this Unit Challenge.*

7.3.4 Reflecting on Your Work

(Key Skill: Personal Reflection)

After your work in Discovery, you should have a better idea of the problems you face and have produced a Problem Statement you could use to tackle the Unit Challenge. Take a moment to reflect on this work. Consider the following prompts but feel free to expand on any to best capture your learning experience and better inform your next steps.

- Of the skills you practiced in Discovery, which were the most challenging? Which were the most interesting?
- How were you most comfortable working during these exercises? In small groups, independently, or with the larger class? Why? How does your choice reflect your personality type and leadership style?
- Was your Problem Statement strictly focused on the environmental problem of sea level rise, or did it also consider important social and economic considerations? How might a focus on the environmental aspects limit your ability to identify truly sustainable solutions?
- You've been given three viable solutions to assess as a part of this case study. But this is not an exhaustive list of options or even necessarily the best possible course of action for every scenario. Take a moment to "think outside the box." Are there any other possible solutions you think would be worth exploring? Describe one that you think would be worth pursuing.

7.4 Climate Migrants: Analysis

Specific Skills You'll Need to Review: Quantitative Literacy, Sustainability Science

Review your Background and Discovery sections before beginning the Rotating Station exercises below. While you focused on one potential solution in your Independent Research in Discovery, keep an open mind as your work through Analysis activities.

7.4.1 Rotating Stations

(Key Skill: Quantitative Literacy)

At each of the following stations, you will review data that are relevant to the three potential solutions you're considering. Spend some time working through the analyses at each station to learn more about this issue and possible solutions for your Unit Challenge.

Be sure to write down one finding at each station that will help inform your selection of a solution.

Station 1: Salt-Tolerant Rice Researchers have been working to breed salt-tolerant varieties of rice that can survive increasingly saline conditions but still maintain nutrition and productivity. To see how various varieties performed at very controlled research stations compared to *in situ* conditions on working farms, Sultana et al. (2019) calculated the following yield gaps for four different locations across coastal Bangladesh (Table 7.1): Yield gap I refers to the difference between research stations and demonstrations plots, while Yield gap II refers to the difference between demonstration plots and actual farms, and total yield gap refers to the difference between the controlled conditions at the research station and real-world conditions on an actual farm

Based on the results of this study, answer the following:

- Calculate the difference in average yield between the research station and demonstration plots at each location (Yield gap I).
- Similarly, calculate the difference in yield between the demonstrations plots and actual farms at each location (Yield gap II).
- Is there a correlation between Yield gap I and Yield gap II? In other words, if a location had a large Yield gap I, was it highly likely it would also have a large Yield gap II?
- Describe the patterns you see.
- What might explain the yield gap between research stations and demonstration plots? Between demonstration plots and actual farms? What does this tell you about how to put research findings into practice on local farms?
- If traditional varieties of rice typically produce 6 t/ha under ideal conditions, on average, what percent of this ideal yield should farmers expect if they switch to the salt-tolerant variety? Why would farmers be interested in using this salt-tolerant variety if it is less productive?

Record any relevant Station 1 findings

Table 7.1 Estimated yield gap of the rice strain Bina dhan-11 at four different locations. (Source: Sultana et al. (2019) {CC BY 4.0})

Parameter	Jamalpur	Mymensingh	Sunamgonj	Sherpu
Average yield of research station (Y_R), $t \cdot ha^{-1}$	5.5	5.60	5.50	5.50
Average yield of demonstration plots (Y_D), $t \cdot ha^{-1}$	4.90	4.50	5.20	4.73
Average yield of actual farm (Y_F), $t \cdot ha^{-1}$	3.99	3.87	4.03	4.08

Station 2: Saltwater Aquaculture Black tiger shrimp (*Penaeus monodon*) are just one of a number of species considered for aquaculture in Bangladesh. Aftabuddin et al. (2021) evaluated black tiger shrimp along with six other candidate fish and shellfish aquaculture species. They looked at potential in terms of environmental, economic, social, and governance factors specific to Bangladesh when constructing the figure below to compare the overall benefits of each species for saltwater aquaculture (mariculture) development.

Based on Fig. 7.9, answer the following:

- Which of the seven candidate species scored the highest in each of the individual domains (economical, environmental, social, governance)?
- How might you approach determining which species performs best overall (highest scores across all four domains)?
- When considering all domains in aggregate, which species scores the lowest? The highest?
- The factors considered in constructing the figures were based on well-established sustainability domains. Give one example of the potential benefit measured for each domain considered (e.g., what are each of these domains actually quantifying in their score?).

Record any relevant Station 2 findings

Station 3: Polder Construction A key factor in determining the long-term effectiveness of any polder in maintaining an agroecosystem in Bangladesh will be the amount of rainfall and evapotranspiration in the area and how that is projected to be affected by climate change.

Mainuddin et al. (2021) evaluated three polders in the Ganges delta region of India and Bangladesh (Fig. 7.10). Expected rainfall and potential evapotranspiration under a variety of climate change scenarios were examined to determine how polders constructed to provide protection based on historical conditions would continue to work under a changing climate. The figure below compares projections for the three polders.

Based on the figure, answer the following:

- Which metric varies most across the three locations (rainfall or PET)? Why might this metric be more variable in the region?
- Assume that if the 95% confidence intervals do not overlap, differences between groups are likely statistically significant. Compare each of the climate scenario models to the historical conditions (gray) at the three locations. Are there any scenarios where rainfall is projected to significantly exceed the historical levels that polders were designed for?
- If CCS2 is the most likely climate scenario for your Unit Challenge location, and this location is most similar to Amtali in characteristics, how confident are you that basing polder construction on historical conditions will provide lasting protection from seawater inundation? Explain your statement.

Record any relevant Station 3 findings

Station 4 In your Unit Challenge you were given a set of assumptions that included estimated costs for implementing each of your three possible solutions, including ongoing yearly costs and expected income that may be gained. Consider that your goal is to identify the net gain/loss for implementing each of these over a 10-year period.

Refer back to the Relevant Facts and Assumptions in the Unit Introduction. For each option, calculate the net revenue to the community with 4 ha of available farmland over a 10-year period.

Record any relevant Station 4 findings

7.4.2 System Mapping

(Key Skill: Sustainability Science)

Your assigned solution may have a variety of direct and indirect economic, social, and ecological impacts that should also be considered. For example, switching to shrimp farming may provide additional economic benefits to a

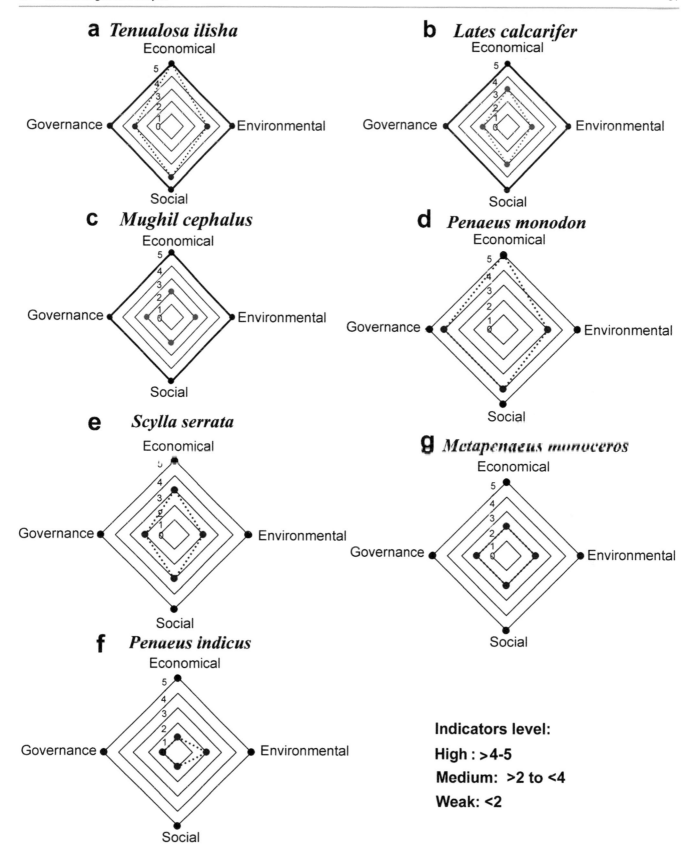

Fig. 7.9 The potential benefit of fish and shellfish species for mariculture in Bangladesh's coastal waterbodies. Potential in each of four domains increases as distance from the center of each figure increases. (Source: Aftabuddin et al. 2021 {Open Access})

Fig. 7.10 Comparison of the annual average rainfall (top) and potential evapotranspiration (PET, bottom) with 95% confidence intervals for the historical period and for five climate change scenarios (CCS1_AELR: average PET and low rainfall, CCS2_AEHR: average PET and high rainfall, CCS3_AEAR: average PET and average rainfall, CCS4_LEHR: low PET and high rainfall, CCS5_HEAR: high PET and average rainfall). (Source: Mainuddin et al. 2021 (CCA 4.0 IL))

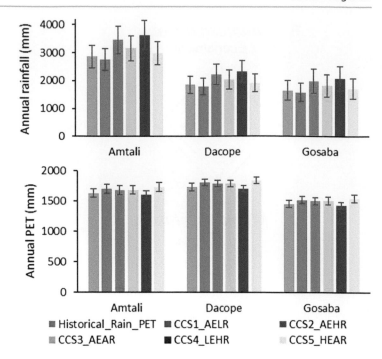

farmer but may degrade the environment and endanger other critical resources like freshwater supplies.

Working with other class members assigned the same solution, develop a simple sustainability map that shows the various system connections across each of the three sustainability domains (ecological, economic, and societal). When you find connections between impacts that cross domains, draw a line to identify the connection (e.g., production of a commercial by-product (economic) leading to increased financial resources for a family (social)).

The goal is to think broadly about this larger system and how implementing actions in one domain (e.g., ecological) may impact components in another domain (e.g., economic or societal).

Use the following template (Fig. 7.11) to get started:

When each group has completed their basic sustainability map, **come together as a class to compare maps for the three possible solutions.** This information will help inform your decision support work later in the unit.

Unit Analysis Summary *Based on your explorations, what have you learned that can help inform your choice of a solution? Do the data support the adoption of one of these potential solutions?*

7.4.3   Reflecting on Your Work

(Key Skill: Personal Reflection)

During your work in Analysis, you explored some of the research into possible solutions to help inform your deci-

sion. Take a moment to reflect on this work. Consider the following prompts but feel free to expand on any to best capture your learning experience and better inform your next steps.

- How did you feel working with data? Do you consider quantitative literacy a strength or an area for improvement for you?
- How important should science be in informing management and policy? Do you feel the data you examined support and justify the costs of addressing your challenge?
- Any solution should be examined using a sustainability science lens. Your goal is to help residents of an agricultural community in coastal Bangladesh survive the effects of rising sea levels. But any solution you implement may have other direct and indirect impacts. What are other possible economic, social, or ecological impacts? How should these considerations influence your decision?

7.5   Climate Migrants: Solutions

Specific Skills You'll Need to Review: Problem Solving, Decision Support, Communicating Science

Review your Unit Challenge and the major findings from Discovery and Analysis, including the sustainability map you created to highlight connections among possible solutions and the larger socioecological system.

Fig. 7.11 Template sustainability map to help connect ecological, societal, and economic considerations associated with your potential solutions

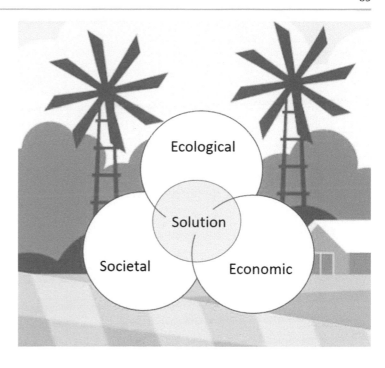

7.5.1   Small Group Guided Worksheets

(Key Skill: Decision Support)

Decision support matrices can help break down the desired outcomes to reflect multiple criteria for consideration, and they allow you to compare how each possible solution achieves those desired outcomes. This not only helps inform decision-making, but it also provides transparency in the decision process and justification to help you advocate for its adoption.

For the three possible solutions, you will evaluate how well each can achieve the following desired outcomes:

- Provides a sustainable food supply for the local community
- Maximizes net profit over a 25-year period
- Represents a lasting solution (not a short-term fix), even with the changing climate
- Is widely acceptable to the community
- Minimizes secondary impacts (e.g., on soil and water quality, on biodiversity of the local ecosystem, etc.)

Considering your three potential solutions, develop a formal decision support matrix to compare and evaluate each approach using the template matrix below (Table 7.2). Getting the most out of the decision matrix requires a depth of knowledge about each of the three possible solutions. Below we list additional sources about each option. Please review each of these, paying particular attention to the two alternatives to your assigned approach.

Note that your group may have uncovered a novel solution not included in this list of three. You may choose to work through the structured decision matrix with your self-identified solution as a fourth solution option.

7.5.2   Additional Sources

1. Salt-tolerant rice: Salt tolerant rice varieties bring hope to Bangladesh farmers |International Centre for Climate Change and Development (ICCCAD)
2. Salt-based agriculture: Shrimp farming in Bangladesh – Responsible Seafood Advocate (globalseafood.org)
3. Polder construction: The polder promise: Unleashing the productive potential in southern Bangladesh | Water, Land and Ecosystems (cgiar.org)

Score each of your three solutions for each of the desired outcomes using a simple relative scale of 1 for least benefit to 5 for greatest benefit. Using a relative scale means you don't need to know exactly how well each solution meets the goal of each desired outcome, but you can use your judgment to assess how well each solution works compared to the others.

While this is a relative (subjective) scale, note that you will need to justify your scoring of each solution for each desired outcome.

Once each cell in your decision matrix has a relative score, calculate an average score for each solution. Based on this analysis, which is the "best" solution considering all your desired outcomes?

Table 7.2 Basic decision matrix to compare the three possible solutions for your Unit Challenge. Include your justification for each solution's ranking for each desired outcome

Solution Options	Desired Outcomes				
	Provides a stable food supply	Maximizes net profit over 10 years	Maximizes duration / longevity of effect	Maximizes stakeholder buy-in	Minimizes secondary impacts
Salt-tolerant rice varieties					
Crop conversion: shrimp farming					
Polder construction					
Justification for ranking					

Use a scale of 1-5 to score each option under each desired outcome category where 1 = does not achieve the desired outcome and 5 = completely achieves the desired outcome.

7.5.3   Role Playing

(Key Skill: Communicating Science)

In tackling environmental issues, you'll often find yourself working with groups of people with different perspectives about implementing a solution. You should use clear, concise communication to summarize your solution and justify its selection. To be effective, this must address concerns likely to be presented by various stakeholder groups and how the risks of taking these actions outweigh the risks of taking no action.

Your group will present and justify your chosen solution to the class with a particular emphasis on how it might benefit or impact one of the key stakeholder groups listed below. Be sure to include arguments that support this solution from the ecological, societal, and economic domains of sustainability science. Listeners will also be **assigned a stakeholder identity**, with an opportunity to **ask follow-up questions** after your presentation that reflect their unique concerns and perspectives. Your job will be to listen carefully and tailor your answers to this audience of stakeholders.

Key stakeholder groups include the following:

- Residents of the community at risk
- A representative of the Bangladesh Ministry of Agriculture
- Member of the Bangladesh Climate Alliance
- A representative of Bangladesh's growing shrimp industry
- A consumer group lobbying for lower rice prices

7.5.4   Reflecting on Your Work

(Key Skill: Personal Reflection)

During your exercises in Solutions, you explored several possible actions that could be taken to help a small agricultural community at risk of flooding in coastal Bangladesh. Take a little more time now to reflect on your findings and the skills you practiced. Consider the following prompts but feel free to expand on any of them to best capture your learning experience and feelings about this issue.

- Reflect on your work today through a sustainability science lens. Does your solution address ecological, economic, and societal considerations? Which considerations do you think should carry the most weight in the decision? Why?
- How did you weigh solutions that might have the greatest environmental impact against those that are most likely to be implemented and maintained for long-term impact?

- How can environmental scientists work to show the value of healthy ecosystems and justify the costs of mitigation strategies?
- Science communication can be challenging, especially when working with diverse audiences. We need to craft our communication to match the interests and values of the target audience, but how do you do this when your audience contains a mix of stakeholder groups? How can you maximize the impact of your message to a diverse audience?

Unit Solution Summary *Summarize and justify your final solution choice and outline how it addresses the direct challenge while also considering social, economic, and ecological impacts. Also demonstrate that it will continue to meet the challenges posed by climate change.*

7.6 Climate Migrants: Final Challenge

As a part of this Unit Challenge, you were asked to write a one-page Fact Sheet summarizing a management plan to help residents of a coastal Bangladesh agricultural community threatened by sea level rise. Your Fact Sheet should include the following components:

- Brief problem statement
- Recommended mitigation strategy with sufficient details to summarize the general approach
- Justification of this recommendation (e.g., long-term effectiveness given anticipated climate changes, implementation costs, other benefits provided, etc.). Be sure to use a sustainability lens to include considerations of direct and indirect ecological, social, and economic considerations
- Any obstacles the group might face trying to implement your solution

Consider using figures, graphics, and tables to help summarize the system and show how this solution is well suited to meet all desired outcomes.

Final Unit Challenge Submit your final recommendations in a one-page Fact Sheet using clear science communication designed for a lay audience.

References

Aftabuddin S, Gulam Hussain M, Abdullah Al M, Failler P, Drakeford BM (2021) On the potential and constraints of mariculture development in Bangladesh. Aquac Int 29:575–593

Church JA, White NJ (2011) Sea level rise from the late 19th to the early 21st century. Surv Geophys 32(4-5):585–602

Environmental Justice Foundation (2018) Climate displacement in Bangladesh. https://ejfoundation.org/reports/climate-displacement-in-bangladesh

Mainuddin M, Karim F, Gaydon DS, Kirby JM (2021) Impact of climate change and management strategies on water and salt balance of the polders and islands in the Ganges delta. Sci Rep 11:7041

Rahmstorf S (2021) Sea level in the IPCC 6th Assessment Report (AR6). Realclimate.org

Rigaud KK, de Sherbinin A, Jones B, Bergmann J, Clement V, Ober K, Schewe J, Adamo S, McCusker B, Heuser S, Midgley A (2018) Groundswell: preparing for internal climate migration. World Bank, Washington, DC. © World Bank

Sultana MR, Rahman MH, Haque MR, Sarkar MA, Islam S (2019) Yield gap of stress tolerant rice varieties Bina dhan-10 & Bina dhan-11 in some selected areas of Bangladesh. Agric Sci 10(11):1438–1452

Core Knowledge

Water resources, Water use and management, Sustainability, Green technologies

8.1 Environmental Issue

In many parts of the world, it has long been a struggle to provide a safe, dependable freshwater supply. Sometimes the problem is water availability, with too many people pulling from too small a resource. But in other cases, the problem is water quality, with an insufficient infrastructure to remove pollutants that make water unsafe for human consumption. Often these issues occur together, with insufficient financial and political support to make the changes necessary to solve the problem. As a result, currently, an estimated 1.6 billion people do not have adequate access to this basic element, with water scarcity impacting public health, food production, economic development, and global security (United Nations 2021).

Over time, water scarcity issues have been exacerbated by a variety of factors. Population growth, particularly in areas with naturally limited freshwater availability (e.g., Las Vegas, NV), threatens groundwater supplies as water extraction rates far exceed replenishment. Irrigation to support increased agricultural production consumes the most fresh water globally (roughly 70%), while the increased use of fertilizers and pesticides also contaminates surface waters.

Economic growth and industrialization often lead to contaminated surface and groundwater supplies. But perhaps of greatest concern is the additional stress that climate change places on this valuable resource. In many parts of the world, droughts are becoming more frequent and more severe, making water scarcity a critical issue for both ecosystems and the human populations they support (Fig. 8.1).

8.2 Background Information

8.2.1 The Problem

Water scarcity occurs when demand for fresh water exceeds supply. Water stress has two components: some areas, like the desert Southwest in the United States, have a physical scarcity of water, while other areas have adequate precipitation but lack the infrastructure to deliver a safe supply to residents. There is sufficient fresh water on Earth to meet the needs of 7 billion people, but water is often wasted, polluted, and, most importantly, distributed unevenly across the globe. While estimates vary widely, experts believe that water scarcity impacts between 1 and 2 billion people globally. Often forced to rely on polluted surface waters, individuals may be at risk from a host of water-borne illnesses like cholera and typhoid fever.

While there are various ways to define and quantify water stress, very few account for water quality or groundwater supply. The most common measurement uses a ratio between physical supply and water demand, but it is important to note that water scarcity is not limited to dry regions. Currently, 43 countries located on every continent suffer from water scarcity (Fig. 8.2). While developing regions face the most widespread water supply concerns, even developed nations are dealing with challenges. For example, while water supply problems in the United States, particularly in drought-prone western states, are nothing new, in recent years, a persistent and increasingly severe drought has led to a declaration of a state of emergency and the implementation of widespread, mandatory water conservation actions in California.

One important factor affecting western US water supplies is the status of the Colorado River, which provides water for 40 million residents in western states. In addition to the Colorado River Basin itself (Fig. 8.3), water from the Colorado is carried to populated areas in Southern California.

J. Pontius, A. McIntosh, *Environmental Problem Solving in an Age of Climate Change*, Springer Textbooks in Earth Sciences, Geography and Environment, https://doi.org/10.1007/978-3-031-48762-0_8

Fig. 8.1 Lake Oroville, CA in 2021 when water levels dropped to 38% of capacity. (Source: Prank Schulenburg [CC BY 3.0] via Wikimedia Commons)

Access to water from the Colorado River follows a pecking order of "first in time, first in right," with the oldest water rights taking priority over others. However, widespread drought and a dwindling water supply led states relying on the Colorado River to negotiate a new plan in 2019 to deal with water shortages that are triggered when water levels in key reservoirs like Lake Mead drop.

Current record low water volumes in the Colorado River and its reservoirs are a result of 3 years of persistent drought. By March 2022, the water level in Lake Powell, a major reservoir on the Colorado, dropped below a critical threshold that threatens the ability of the Glen Canyon dam to generate power. By the end of April 2022, water levels in Lake Mead fell to 321.5 m, a level not seen since 1937, leaving high and dry one of the original water supply intakes on the floor of the reservoir.

Adding to California's water woes, hotter temperatures and dry conditions have resulted in moderate to severe drought conditions for much of the state. Its own network of water supply reservoirs is also at risk. An interactive map of water levels (Fig. 8.4) shows updated conditions in 12 major water supply reservoirs in California and highlights the reduced capacity in systems like Shasta and Trinity.

Faced with diminishing supplies (estimated to be able to provide 302 L per Californian/day in April 2022) from both the Colorado River system and its own reservoirs and unable to continue to meet the historic demands of 473 L per person/day, California water managers declared a State of Emergency with some local residents put under emergency mandatory water use restrictions. The pressure to conserve existing supplies and develop new ones will likely only increase over time in drought-stricken areas.

8.2.2 The Role of Climate Change

Few issues are more directly linked to climate change than water supply availability. Not only will the warmer, drier conditions expected to occur in areas like the western United States lead to increased evaporation and reduced surface water levels, but other, more subtle, impacts will add to the crisis. For example, mountain snowmelt, a critical source of water for rivers like the Colorado, will occur earlier, exposing soils to direct sunlight and drying them out earlier in the year. These drier soils will absorb more rainfall, further reducing the amount of precipitation flowing into rivers.

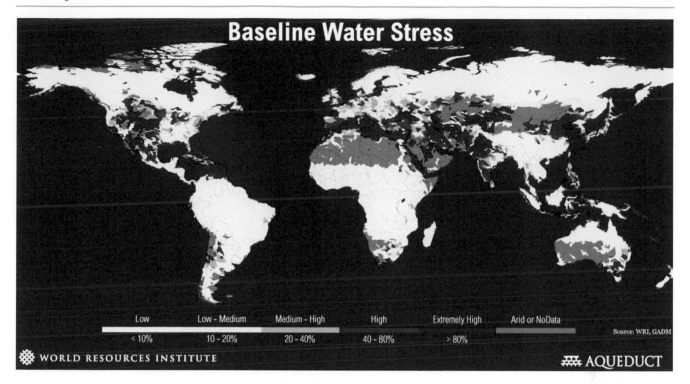

Fig. 8.2 A ratio of total annual water withdrawals to total available annual renewable supply highlights various levels of water stress. (Source: World Resources Institute [CC BY SA 4.0] via Wikimedia Commons)

Warmer temperatures also mean less snowfall, which leaves less water available in local reservoirs later in the season when water demand is highest.

Even sea level rise expected with climate change will threaten water supplies in California. Not only will saltwater intrusion threaten coastal groundwater supplies, but a rising sea level will threaten the Sacramento-San Joaquin Delta, a critical part of the California water supply system and the source of water for 25 million residents. To prevent saltwater intrusion, more fresh water will need to be passed through the delta, reducing the amount available for people to consume. Adding to California's water supply vulnerability as the climate changes is the state's water infrastructure. Most of the state's roughly 1,500 dams and reservoirs are at least a half century old, and they were not designed to handle the extreme events expected under a changing climate (Sahagun 2021). Major dams are projected to be five times more likely to flood this century than last, increasing the risk of catastrophic failures.

According to Sommer (2021), about half the loss of water from the Colorado River is due to warmer temperatures, and every increase of 1°C will result in an additional 9% decrease in river flow. Because of this connection to temperature, hydrologists project that California's overall water supply, which serves 25 million residents through two state and federal water projects, will continue to shrink as temperatures increase even if the state gets more rain than average.

Adding to the problem is that warmer temperatures will likely increase demand for additional supplies (e.g., increased need for irrigation as agricultural soils dry out more rapidly) at the very time available water is decreasing. As supplies dwindle and demand likely increases, intrastate conflicts over diminishing water supplies will likely worsen.

8.2.3 Solutions

Faced with the prospect of diminished availability of traditional sources of water for potable use, cities in the impacted areas of the western Unites States are focusing on both sides of the supply and demand water scarcity equation by working to maximize water conservation efforts and develop alternate sources of freshwater supply.

Reducing Water Demand The western United States has had extensive experience conserving available water supplies. In many areas, new homes being built are routinely required to install low-flow toilets and other fixtures, utilize natural desert landscaping, use drip irrigation and gray water reclamation, and capture rain water for domestic use.

At the municipal level, towns have used xeriscaping in public areas and restricted such activities as car washing and watering lawns. Additional steps have included public education efforts (Fig. 8.5) as well as economic incentives like

Fig. 8.3 The Colorado River Basin. (Source: USGS {Public Domain})

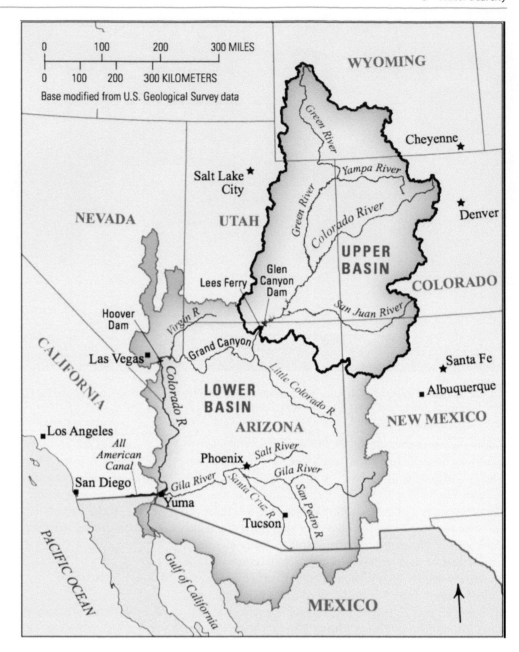

setting water rates based on usage in order to encourage conservation (use more, pay more). In reality, many of the easiest water conservation steps that can be mandated are already in place, and additional water conservation steps will not likely bridge the gap between supply and demand.

Increasing Water Supply Where fresh water is limited, there are several alternative sources of potable water for communities with sufficient financial resources.

Desalination: Removing salt from ocean water to produce potable fresh water has become a commonly used approach in areas with limited freshwater supplies (Fig. 8.6). As of 2022, there were about 17,000 desalination plants globally

and 1,400 in the Unites States, with the majority of US plants desalting brackish groundwater. The largest plant in the Unites States, constructed on the shores of the Pacific Ocean in Carlsbad, CA, uses reverse osmosis to produce 50 million gallons per day (mgd) of freshwater. While desalination can provide local, reliable potable water, it is expensive and typically meets only a fraction of overall demand.

Potable water reuse: It is possible to treat municipal wastewater to such a degree that it can be used for drinking. After it goes through the treatment plant, wastewater is processed in a second plant (and sometimes a third) using advanced chemical, biological, and physical treatments. The water is then added directly into the drinking supply system or stored in a natural system (rivers, lakes, aquifers, or reser-

Storage in Selected Major Reservoirs

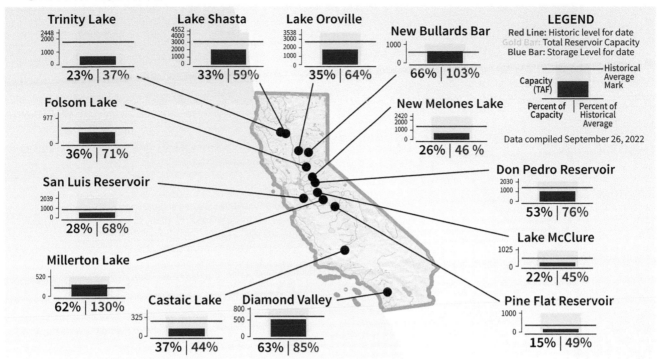

Fig. 8.4 California's interactive map of water levels in major reservoirs shows current conditions (Shown here for June 14, 2022). (Source: CA Dept. of Water Resources (Public Domain))

voirs) (Fig. 8.7). In the latter scenario, water is subsequently extracted from the natural system, treated again, and then distributed to people for drinking or other uses. In both cases, the resulting water is termed reused.

New water supply reservoirs: While much of the focus in California has been on *large dams* and reservoir projects, like the proposed 1.8 million acre-foot Sites Reservoir to be built at a cost of about $5 billion, the state has a number of very small water supply reservoirs. The French Lake Reservoir in Nevada County, CA, has a storage capacity of 55 acre feet, while the Rowena Reservoir in Los Angeles has a capacity of only 30 acre-feet. These smaller water storage basins help maintain water supply throughout dry summer months and reduce the need to transport water over long distances, which often results in water loss due to leaks.

Underground storage: https://www.epa.gov/uic/aquifer-recharge-and-aquifer-storage-andrecovery#:~:text=The%20objective%20of%20AR%20is,aquifers%20and%20control%20land%20subsidence. Aquifers supply high quality potable water to many parts of the western Unites States. Unlike surface water, aquifers have not historically been regulated, sometimes leading to significant depletion of this critical resource. Groundwater levels are now monitored and managed, with current conditions published by the California Department of Water Resources (Fig. 8.8). To enhance the overall underground water supply, California has been injecting both treated wastewater and stormwater runoff into aquifers.

8.2.4 Unit Challenge

San Crucible, a hypothetical small community of 3000 single-family homes that lies along the Pacific coastline between San Diego and Los Angeles, has relied on water transported from the Colorado River for half of its domestic water supply. Town managers have been told that there will be no more water available from the Colorado beginning on June 30, 2025. The overarching challenge you face is to determine how to best help this small community meet its freshwater needs, without impacting its economic stability. Specifically, **your Unit Challenge is to identify how to meet current water demand without access to the Colorado River.** However, your plan must also consider the long-term needs of the community as it continues to grow, as well as the impacts of ongoing climate change.

8.2.5 The Scenario

The town has hired you, an environmental scientist with a strong background in western water issues, to help them solve their water supply problems. Your initial hope was that mandating additional water conservation steps throughout the community might simplify your task, but you discover

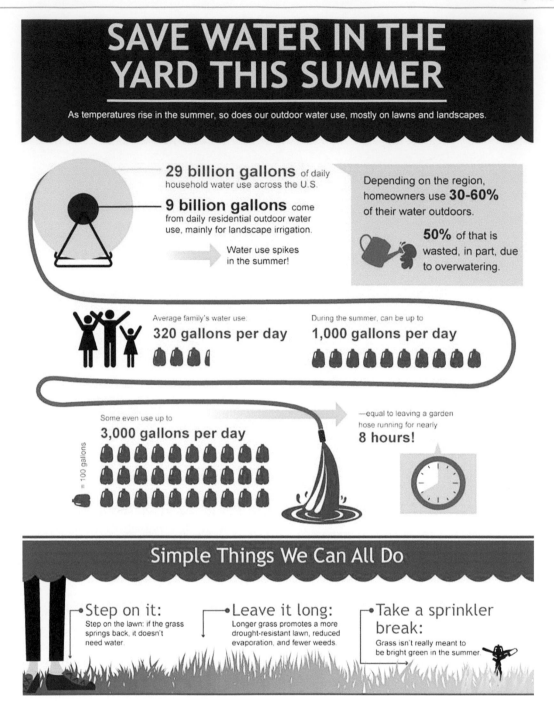

Fig. 8.5 The WaterSense program works to educate the public about ways to protect the nation's water supply. (Source: US EPA {Public Domain})

that all the homes already practice state-of-the-art water conservation.

You'll have to find new sources of water to meet the town's needs.

After reviewing all the possible options for the town, you decide to focus on three:

1. **Water supply reservoir**. The nearby free-flowing Crucible River could be dammed to provide a freshwater supply for the residents. Because of the hydrology of the watershed, the dam could be constructed at a reasonable cost, but environmental concerns include impacts on habitats of fish and other aquatic life in the river, increased water temperatures and reduced dissolved oxygen in waters below the dam, property owner rights on impacted land, and vulnerability to drought in a climate-driven future.

Specific recommendation: Construct a small dam along the Crucible River to create a 250 acre-foot reservoir

Water desalination

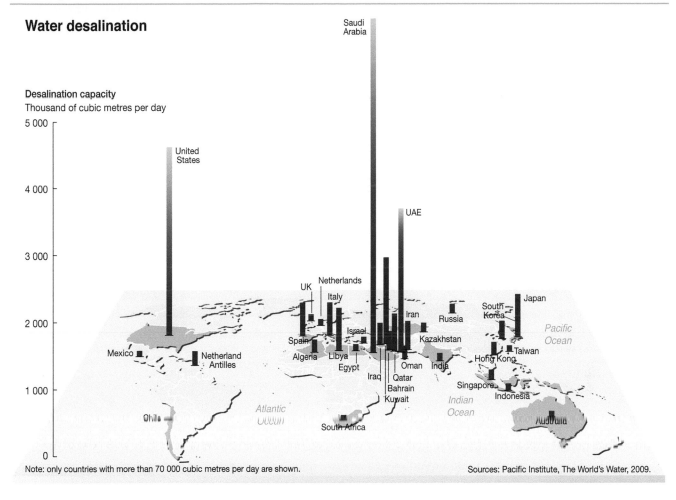

Desalination capacity
Thousand of cubic metres per day

Note: only countries with more than 70 000 cubic metres per day are shown.

Sources: Pacific Institute, The World's Water, 2009.

Fig. 8.6 Desalination is an increasingly important practice to secure clean water in a number of countries. (Source: GRID-Arendal via Flickr www.grida.no/resources/7609)

to provide potable water for the community throughout the year. Limited recreational use would be permitted for town residents (e.g., non-motorized boating).

2. **Desalination plant**. Removing salt from sea water to provide a source of potable water is a well-established technology that can take a sustainable resource (salt water) and safely convert it to a scarce one (fresh water). While costs are high, this option would also provide additional jobs for the community. Environmental concerns about desalination include the amount of electricity used during the process, damage to marine life at ocean water intakes, and brine waste disposal issues.

   Specific recommendation: Construct and operate a small desalination plant to provide potable water for the community.

3. **Water reuse.** Effluents from the town's small wastewater treatment plant could be treated in a series of well-established cost-effective steps to provide high-quality potable water for community residents. A major concern is the public perception of water reuse (e.g., the original phrase "toilet-to-tap" did little to improve the public's acceptance of recycled wastewater). Specific concerns

center on the possible presence of harmful chemicals and pathogens in the recycled water.

   Specific recommendation: Invest in the necessary technologies at the community's existing wastewater treatment plant to produce enough potable reused water to meet the town's needs.

Each of these three approaches has important advantages and disadvantages and costs and benefits. Your task is to evaluate each and recommend the approach that will, in your opinion, provide the most fresh water at the lowest price, while minimizing impacts on the local ecology.

8.2.6   Relevant Facts and Assumptions

- Each home in the 3000-home community uses an average of 400 gallons of water daily.
- The town will need to replace the 50% of its total daily usage previously extracted from the Colorado River to meet its needs.
- The town population is growing at a rate of 1% each year.

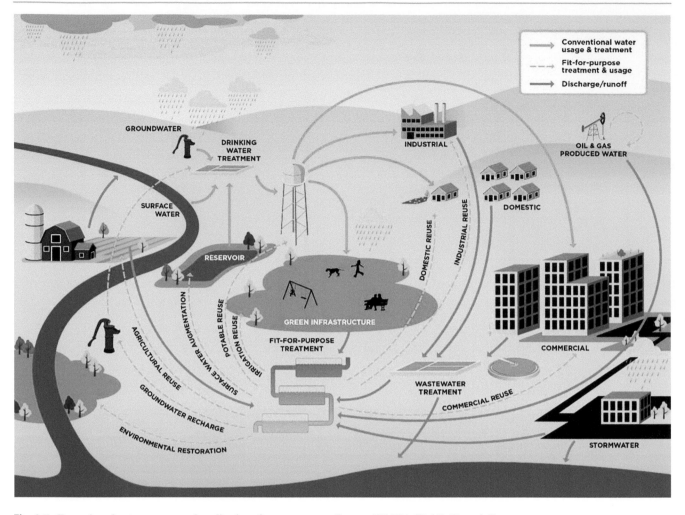

Fig. 8.7 Examples of water sources and applications for water reuse. (Source: US EPA {Public Domain})

• The future climate of San Crucible is expected to be hotter and drier.
• Below are additional assumptions for each solution:

  – **Water supply reservoir**: A dam could be constructed to create a 250 acre-foot reservoir of fresh water. The average cost for constructing a reservoir in CA is $1560 per acre-foot with additional annual costs of $5000 for maintenance and basic water treatment. One acre-foot of water provides 325,851 gallons of drinking water, and the river can be expected to recharge 25 acre-feet monthly.
  – **Desalination plant**: Construction of a desalination plant costs $475,000 per mgd of potable water production capacity and an additional $75,000 per mgd each year to maintain and operate. Assume that your limit is a plant capable of providing 1 mgd.
  – **Water reuse:** Construction of a water reuse plant costs $225,000 per mgd. The average cost for treating wastewater for water reuse at this plant would be $50,000/

mgd annually. This plant could treat 60% of the 1 mgd of water entering the system.

8.2.7 Build Your Foundational Knowledge

Below are web sources that provide additional information about each of the solutions you're considering for this Unit Challenge. This information provides a critical foundation to help you evaluate each option and support your final choice. After reviewing each source, be prepared to answer questions in the Unit Preparation Assessment Quiz and to summarize any information relevant to your Unit Challenge.

Water supply reservoir:
Human-made Reservoirs

Desalination plant:
Thirsty? How 'bout a cool, refreshing cup of sea water?

Fig. 8.8 Map of 2022 groundwater levels from California's Groundwater Livesite. (Source: CA Department of Water Resources {Public Domain})

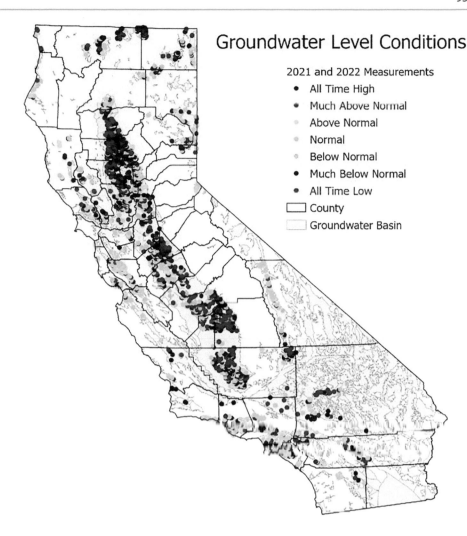

Water reuse:

Potable Reuse 101: An innovative and sustainable water supply solution

Final Product: A one-page Fact Sheet summarizing the issue, detailing your solution, and justifying your choice of that solution. Consider your audience, town residents and officials, and be sure to demonstrate how your proposed solution will stand up to the challenges posed by climate change.

8.2.8 Preparation Assessment Quiz

Are you ready to tackle your challenge? At this point you should understand the basic environmental principles and ecological processes involved in this environmental problem.

Consider the following questions. If you are comfortable with answering these, then you are ready to head into Discovery, Analysis, and Solutions activities.

- How do scientists assess water scarcity?
- What are the primary causes of water scarcity globally?
- What are three ways that climate change can impact water supply?
- How might sea level rise impact supplies of freshwater?
- According to the link Human-made Reservoirs, what are some of the pros and cons of artificial lakes?
- According to this USGS desalination article, in 2015, how many desalination plants were there in the world and how much potable water did they produce?
- According to the American Water Works Association article, what is the difference between direct and indirect water reuse?
- For each of the proposed solutions, are there any additional benefits that might arise from their implementation that might not be directly related to your Unit Challenge?
- For each of the proposed solutions, are there any negative unintended consequences that might result from their implementation?
- What additional information did you glean from your web sources that might help inform your Unit Challenge?

8.3   Water Scarcity: Discovery

Specific Skills You'll Need to Review: Navigating the Scientific Literature, Science Communication, Problem-Solving

8.3.1   Independent Research

(Key Skill: Navigating the Scientific Literature)

To better understand the various approaches being considered to meet the freshwater needs of San Crucible, you first need to examine the literature to see what others have found. Conduct a search of the peer-reviewed scientific literature focused on the solution you have been assigned and identify one research paper that focuses on your assigned approach.

Prepare a summary of the article you selected that includes the following:

- **Citation**
- **Main topic**: Stick to a few words, likely pulled from the title.
- **General summary**: A few bulleted sentences summarizing the research question it addresses and approaches it takes.
- **Methods:** How did they approach their research question?
- **Location**: Where was the work done?
- **Conclusions**: Concise list of the findings, specifically capturing the take-home message.
- **Relevance**: How might this study help inform your Unit Challenge? Feel free to make a bulleted list of points you may want to include later.

8.3.2   Literature Share: Reciprocal Instruction

(Key Skill: Scientific Communication)

Share In small groups, share and critique the research article you found. Keep in mind that your peers have not read this article, and it is your job to convey the key information to them. Note the items that will be important to consider when you are developing your solution to the challenge.

Critique Evaluate how these studies might help inform your Unit Challenge. Consider the following:

- Source (Quality of the work or bias of the authors)

- Methods (Did their methods sufficiently address the research question?)
- Conclusions (Did the results justify the conclusions made?)
- Relevance (Can these findings be applied to your challenge?)

Based on your critiques, choose one article to share with the larger class, along with the key information that may be useful in deciding on a solution to propose.

8.3.3   Think-Compare-Share

(Key Skill: Problem Solving)

Now that you have more information about possible solutions for this unit's challenge, you need to **develop a more formal problem definition** to guide your work throughout the rest of the exercises.

Think Start by working independently to develop a specific Problem Statement to guide the remainder of your work. Problem Statements provide the relevant information and boundaries to make the issue something you can effectively assess and tackle. The basics of a formal Problem Statement include the following:

Problem Statement: A short, concise statement summarizing the issue that includes:

- A **description** of the undesired condition or change that you hope to achieve (What is the actual problem?)
- **Justification** for addressing the problem (Why does this problem matter?)
- Potential **sources** or **causes** of this problem (What is the cause you need to address?)
- The **metrics** you will use to assess the status of the problem (How will you know if you are making a difference?)
- The **desired outcome** for these metrics (What is the end goal or ideal state?)
- Potential **solutions** to consider (How might you attempt to achieve this goal?)

Compare/Share Now return to your small group to share your Problem Statements. Use each of your ideas to develop a joint Problem Statement that contains all key information and is concise, clear, and well written.

Unit Discovery Summary   *Submit a final Problem Statement that succinctly captures the key information to guide your work on this Unit Challenge.*

8.3.4   Reflecting on Your Work

(Key Skill: Personal Reflection)

After your work in Discovery, you should have a better idea of the problems you face and have produced a Problem Statement you could use to tackle the Unit Challenge. Take a moment to reflect on this work. Consider the following prompts but feel free to expand on any to best capture your learning experience and better inform your next steps.

- Of the skills you practiced in Discovery, which were the most challenging? Which was the most interesting?
- How were you most comfortable working during these exercises? In small groups, independently, or with the larger class? Why? How does your choice reflect your personality type and leadership style?
- Was your Problem Statement strictly focused on the environmental problem of water scarcity, or did it also consider important social and economic considerations? How might a focus on the environmental aspects limit your ability to identify truly sustainable solutions?
- You've been given three viable solutions to assess as a part of this case study. But this is not an exhaustive list of options or even necessarily the best possible course of action for every scenario. Take a moment to "think outside the box." Are there any other possible solutions you think would be worth exploring? Describe one that you think would be worth pursuing.

8.4   Water Scarcity: Analysis

Specific Skills You'll Need to Review: Quantitative Literacy, Sustainability Science

Review your Background and Discovery sections before beginning the Rotating Station exercises below. While you focused on one potential solution in your Independent Research in Discovery, keep an open mind as your work through Analysis activities.

8.4.1   Rotating Stations

(Key Skill: Quantitative Literacy)

At each of the following stations, you will review data that are relevant to the three potential solutions you're considering. Spend some time working through the analyses at each station to learn more about this issue and possible solutions for your Unit Challenge.

Be sure to write down one finding at each station that will help inform your selection of a solution.

Station 1: Water Reuse   A major concern with using treated wastewater for potable purposes is public acceptability. Redman et al. (2019) studied differences in public perceptions of potable water reuse for various applications among residents of northern Nevada (Fig. 8.9). The authors surveyed opinions about reclaimed water among urban, suburban, and rural residents.

Considering the patterns shown in the above Likert scale diagram, answer the following:

- How did support for water reuse differ among urban, suburban, and rural communities?
- For which specific use was support among suburban residents the strongest? The weakest?
- Considering that scores of both 4 and 5 indicate support for the application of reused water, which set of applications would suburban residents like those in San Crucible support?
- Considering the estimates of water use across applications (in red) and your list of supported applications above, what proportion of total water demand could be met with water reuse for supported applications in suburban communities?

Record any relevant Station 1 findings

Station 2: Desalination   Discharge of brine wastewater from desalination plants is among the environmental concerns associated with the technology. Saeed et al. (2022) assessed the impacts of brine discharge from the world's largest desalination plant on the coastline of Saudi Arabia (Table 8.1). Among the approaches they used were measurements of bacterial luminescence, a commonly used method for detecting the presence of contaminants and measuring pollutant toxicity in marine waters. This method assumes that bioluminescence is inhibited when water quality is poor. In addition, they measured a number of trace metals at the site.

Based on the data in the table above, answer the following:

- Where was inhibition of bacterial bioluminescence significantly different from the open sea control?

Fig. 8.9 Average Likert scale responses for support for reclaimed water use applications by residential location type. 1 = Strongly Oppose, 2 = Oppose, 3 = Neutral, 4 = Support, 5 = Strongly Support. Estimated proportion of total water use for each application in red. (Source: Modified from Redman et al. 2019 (Open Access))

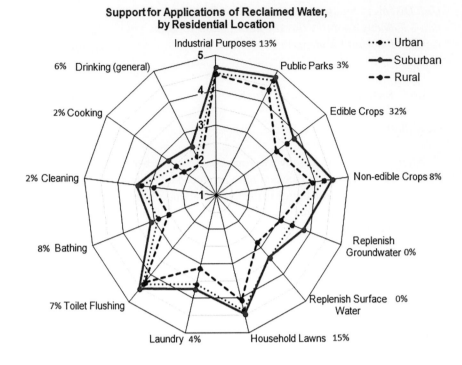

Table 8.1 Inhibition of bacterial bioluminescence in different water sources and mean concentration of trace metals in water (ug/L) at the Jubail Desalination and Power Plants

A. Water source	% Inhibition	Fe	Cu	Ni	Cr
Open sea (control)	8.6 ± 5.1	20.2[a]	BDL	BDL	BDL
Brine discharge (main channel)	*21.7 ± 9.1	22.6[a]	5.8[a]	BDL	BDL
Brine discharge (to open water)	10.9 ± 6.2	26.0[b]	3.5[b]	BDL	1.5
Sediment wash discharge (to open water)	9.8 ± 5.6	166[c]	7.7[c]	BDL	BDL

Source: Modified from Saeed et al. (2022). (CCA 4.0 IL)
*Significantly different from control.[a, b, c, d, e] Metal means with same letter superscript are similar (Analysis of Variance and paired t-tests; $p < 0.05$)
BDL: below detection limit

- What do the data in the table tell you about the severity of the impacts of brine discharge from desalination facilities?
- Which of the metal contaminants measured at the three discharge sites were higher than the control?
- Of the metal contaminants you identified in the preceding answer, which would most likely pose a threat to marine life? Why?
- Based on a quick review of the marine literature, which group of marine biota would you expect to be most sensitive to this contaminant?

Record any relevant Station 2 findings

Station 3: Water Supply Reservoir Dam construction can have many benefits in addition to the creation of a water supply storage system. For example, many dams double as hydropower stations to generate sustainable electricity. But dam construction can also have significant negative impacts on the ecosystem, with far-reaching

effects. Angarita et al. (2018) examined how migratory fish habitats could be affected as more hydropower dams come online in the Mompos Depression, the largest wetland in Columbia.

Figure 8.10 shows various indicators of the degree of alteration to river systems and fish habitats across the Mompos Depression as more hydropower dams come online over time. The black line shows actual historical measurements; colored dots indicate various modeled conditions as more dams are constructed; and the shaded vertical bar highlights the expected maximum dam capacity in the basin.

- Describe the changes in connected spawning habitat and migratory fish habitats currently witnessed based on historical measurements.
- What might account for differences in the various modeled expansion scenarios (different colored dot trajectories) beyond the current 10,000 MW basin capacity?

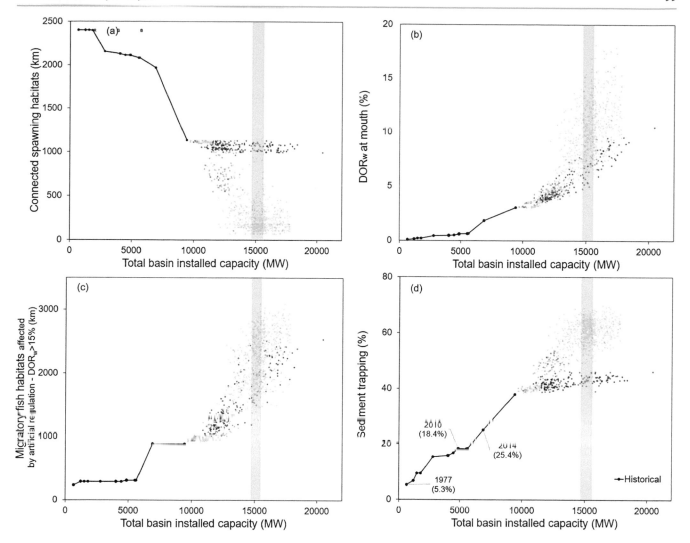

Fig. 8.10 Indicators of basin level cumulative impacts of dam construction under randomly generated dam expansion scenarios. (**a**) Longitudinally connected migratory fish spawning habitat (river length), (**b**) cumulative streamflow regulation measured as weighted degree of regulation (DORw), (**c**) amount of migratory fish habitat (river length) affected by artificial regulation (DOR>15%), (**d**) total sediment trapping in reservoirs upstream of the Mompós Depression. (Source: Angarita et al. 2018 (CC BY 4.0))

- As an environmental scientist adhering to the Precautionary Principle, which modeled scenario (or set of scenarios represented by different colors) would you recommend water managers use to guide decisions about dam construction? Justify your answer.
- Consider that migratory fish in this basin need 1000 km of connected spawning habitats (8.10a) and fewer than 1500 km of affected migratory fish habitat (8.10c) to maintain their populations over time. Estimate what the maximum basin installed capacity would need to be to remain within these limits under the most conservative scenario.

Record any relevant Station 3 findings

Station 4 For your Unit Challenge, you were given a set of assumptions and relevant facts to help guide your decisions.

Remember that you need to replace the 50% of the community's annual water supply. Using this information, calculate the following:

- How many gallons of drinking water could the new water supply reservoir be expected to sustainably provide each year (based on a recharge rate of 25 acre-feet monthly)?
- What proportion of the annual water needs in the 3000-household community would each solution supply (assume the average home uses 400 gallons of water daily)?
- How much would it cost to develop and maintain each of these solutions over a 10-year period?
- What is the cost per gallon provided for each of these solutions at the end of the 10-year period?

Record any relevant Station 4 findings

8.4.2 System Mapping

(Key Skill: Sustainability Science)

Your assigned solution may have a variety of direct and indirect economic, social, and ecological impacts that should also be considered. For example, while construction of a water supply reservoir, even a small one, might have important effects on the river (ecological impacts), it might also provide some limited recreational opportunities for residents (social impact).

Working with other class members assigned the same solution, develop a simple sustainability map that shows the various system connections across each of the three sustainability domains (ecological, economic, and societal). When you find connections between impacts that cross domains, draw a line to identify the connection (e.g., employment opportunities for town residents (economic) leading to increased financial resources for a family (social)).

The goal is to think broadly about this larger system and envision how implementing actions in one domain (e.g., ecological) may impact components in another domain (e.g., economic or societal).

Use the following template (Fig. 8.11) to get started:

When each group has completed their basic sustainability map, **come together as a class to compare maps for the three possible solutions.** This information will help inform your decision support work later in the unit.

Unit Analysis Summary *Based on your explorations, what have you learned that can help inform your choice of a solution? Do the data support the adoption of one or more than one of these potential solutions?*

Fig. 8.11 Template sustainability map to help connect ecological, societal, and economic considerations associated with your potential solutions

8.4.3 Reflecting on Your Work

(Key Skill: Personal Reflection)

During your work in Analysis, you explored some of the research into possible solutions to help inform your decision. Take a moment to reflect on this work. Consider the following prompts but feel free to expand on any to best capture your learning experience and better inform your next steps.

- How did you feel working with data? Do you consider quantitative literacy a strength or an area for improvement for you?
- How important should science be in informing management and policy? Do you feel the data you examined support and justify the costs of addressing your challenge?
- Any solution should be examined using a sustainability science lens. Your goal is to find an alternate water supply for the residents of San Crucible. But any solution you implement may have other direct and indirect impacts. What are other possible economic, social, or ecological impacts? How should these considerations influence your decision?

8.5 Water Scarcity: Solutions

Specific Skills You'll Need to Review: Problem-Solving, Decision Support, Communicating Science

Review your Unit Challenge and the major findings from Discovery and Analysis, including the sustainability map you created to highlight connections among possible solutions and the larger socioecological system.

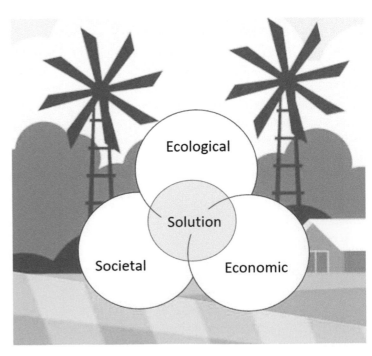

8.5.1   Small Group Guided Worksheets

(Key Skill: Decision Support)

Decision support matrices can help break down the desired outcomes to reflect multiple criteria for consideration, and they allow you to compare how each possible solution achieves those desired outcomes. This not only helps inform decision-making, it also provides transparency in the decision process and justification to help you advocate for adopting a particular solution.

For the three possible solutions, you will evaluate how well each can achieve the following desired outcomes:

- Provides a reliable water supply
- Minimizes costs
- Provides a lasting impact (not a short-term fix), even with a changing climate
- Is widely acceptable to stakeholder groups
- Minimizes secondary impacts (e.g., economic, environmental)

Considering your three potential solutions, develop a formal decision support matrix to compare and evaluate each approach using the template matrix below (Table 8.2). Getting the most out of the decision matrix requires a depth of knowledge about each of the three possible solutions. Below we list additional sources about each option. Please review each of these, paying particular attention to the two alternatives to your assigned approach.

Note that your group may have uncovered a novel solution not included in this list of three. You may choose to work through the structured decision matrix with your self-identified solution as a fourth solution option.

8.5.2   Additional Sources

1. Water supply reservoir: Small water reservoirs – Sources of water or problems?
2. Desalination plant: Desalination | U.S. Geological Survey (usgs.gov)
3. Potable water reuse: 2017 Potable Reuse Compendium (epa.gov) (Chapter 1)

Score each of your three potential solutions for each of the desired outcomes using a simple relative scale of 1 for least benefit to 5 for greatest benefit. Using a relative scale means, you don't need to know exactly how well each solution meets the goal of each desired outcome, but you can use your judgment to assess how well each solution works compared to the others.

While this is a relative (subjective) scale, note that you will need to justify your scoring of each solution for each desired outcome.

Table 8.2 Basic decision matrix to compare the three possible solutions for your Unit Challenge. Include your justification for each solution's ranking for each desired outcome

	Desired Outcomes				
Solution Options	**Provides a reliable water supply**	**Minimizes costs**	**Maximizes longevity of effect**	**Maximizes stakeholder buy-in**	**Minimizes secondary impacts**
Water supply reservoir					
Desalination plant					
Water reuse					
Justification for Ranking					

Use a scale of 1-5 to score each option under each desired outcome category where 1 = does not achieve the desired outcome and 5 = completely achieves the desired outcome.

Once each cell in your decision matrix has a relative score, calculate an average score for each solution. Based on this analysis, which is the "best" solution, considering all your desired outcomes?

8.5.3   Role Playing

(Key Skill: Communicating Science)

In tackling environmental issues, you'll often find yourself working with groups of people with different perspectives about implementing a solution. You should use clear, concise communication to summarize your solution and justify its selection. To be effective, you must address concerns likely to be presented by various stakeholder groups and how the risks of taking these actions outweigh the risks of taking no action.

Your group will present and justify your chosen solution to the class with a particular emphasis on how it might benefit or impact one of the key stakeholder groups listed below. Be sure to include arguments that support this solution from the ecological, societal, and economic domains of sustainability science. Listeners will also be **assigned a stakeholder identity**, with an opportunity to **ask follow-up questions** after your presentation that reflect their unique concerns and perspectives. Your job will be to listen carefully and tailor your answers to this audience of stakeholders.

Key stakeholder groups include the following:

- Local residents who will be served by the new water supply
- State regulators responsible for maintaining water quality
- Engineering firms interested in bidding on the construction of any facilities
- Representatives of California's Sierra Club

8.5.4   Reflecting on Your Work

(Key Skill: Personal Reflection)

During your work in Solutions, you have explored several possible actions that could be taken to provide the residents of San Crucible a reliable water supply. Take a little more time now to reflect on your findings and the skills you practiced. Consider the following prompts but feel free to expand on any of them to best capture your learning experience and feelings about this issue.

- Reflect on your work through a sustainability science lens. Does your solution address ecological, economic, and societal considerations? Which considerations do you think should carry the most weight in the decision? Why?
- How did you weigh solutions that might have the greatest environmental impact against those that are most likely to be implemented and maintained for long-term impact?
- How can environmental scientists work to show the value of healthy ecosystems and justify the costs of mitigation strategies?
- Science communication can be challenging, especially when working with diverse audiences. We need to craft our communication to match the interests and values of the target audience, but how do you do this when your audience contains a mix of stakeholder groups? How can you maximize the impact of your message to a diverse audience?

Unit Solution Summary *Summarize and justify your final solution choice and outline how it addresses the direct challenge while also considering social, economic, and ecological impacts. Also demonstrate that it will continue to meet the challenges posed by climate change.*

8.6   Water Scarcity: Final Challenge

As a part of this Unit Challenge, you were asked to write a one-page Fact Sheet justifying your choice for an approach to supply potable water to the San Crucible community. Your Fact Sheet should include the following components:

- Brief problem statement
- Recommended mitigation strategy with sufficient details to summarize the general approach
- Justification of this recommendation (e.g., long-term effectiveness, given anticipated climate changes, implementation costs, other benefits provided, etc.). Be sure to use a sustainability lens to include considerations of direct and indirect ecological, social, and economic considerations
- Any obstacles the group might face trying to implement your solution

Consider the use of figures, graphics, and tables to help summarize the system and how this solution is well suited to meet all desired outcomes.

Final Unit Challenge Submit your final recommendations in a one-page Fact Sheet using clear science communication designed for a lay audience.

References

Angarita H, Wickel AJ, Sieber J, Chavarro J, Maldonado-Ocampo JA, Herrera-R GA, Delgado J, Purkey D (2018) Basin-scale impacts of hydropower development on the Mompós Depression wetlands, Colombia. Hydrol Earth Syst Sci 22:2839 2865

Redman S, Ormerod KJ, Kelley S (2019) Reclaiming suburbia: differences in local identity and public perceptions of potable water reuse. Sustainability 11(3):564

Saeed MO, Ershath MM, Barnawi AM (2022) Assessing the toxic effects of brine discharge from the world's largest desalination plant, Gulf Coast of Saudi Arabia. Membrane Sci Int 1(1):3–11

Sahagun L (2021) California's aging dams face new perils, 50 years after Sylmar quake crisis. Los Angeles Times. Climate & Environment

Sommer L (2021) The drought in the western U.S. Is getting bad. Climate change is making it worse. National Public Radio. Environment. https://www.npr.org/2021/06/09/1003424717

United Nations (2021) Billions risk being without access to water and sanitation services by 2030. UN News. Global Perspective-Human Stories. https://news.un.org/en/stories/2021/07/1095202

Core Knowledge

Air quality, Atmospheric science, Human health, Energy and transportation

9.1 Environmental Issue

Poor air quality harms both environmental quality and human health. Nowhere are these impacts more apparent than in urban areas, where a perfect storm of vehicle exhaust, emissions from power generation, industrial pollution, and emissions from fuel used to heat homes forms in areas where human populations are high (Fig. 9.1). Often, it is the most vulnerable populations which suffer from poor air quality, resulting in increased risk of stroke, heart disease, lung cancer, and chronic and acute respiratory diseases like asthma. In this unit, you'll explore the connections between climate change and air quality in urban areas.

9.2 Background Information

9.2.1 The Problem

Transportation is one of the largest sources of greenhouse gases (GHGs) as well as a mixture of other hazardous pollutants like ozone, nitrogen and sulfur oxides, and volatile organic compounds (VOCs). According to the Massachusetts Department of Environmental Protection (2022), the vehicles we drive contribute 40% of the nation's smog-causing pollutants and more than 25% of its GHG emissions.

Transportation-related air pollution is concentrated in cities, with both large numbers of residents and commuters from surrounding areas coming in to urban centers to work. While many cities have made impressive strides in developing mass transit systems, motor vehicles are still the leading source of transportation-related urban air pollution, including both regulated air pollutants like ozone and GHGs.

As the climate continues to warm, urban air quality is expected to worsen, as warmer temperatures create more favorable conditions for the formation of ground-level ozone. While steps like cleaner mass transit systems and car-pooling will help reduce the problem, additional action will be needed to provide relief.

More than half the world's population live in cities, and this figure may increase to two thirds by 2030 (United Nations 2019). With large numbers of people living in close proximity to a variety of air pollution sources, it is no surprise that major urban areas grapple with poor air quality, whether it's brown smog largely attributable to Los Angeles traffic or gray smog formed by emissions from China's coal-fired power plants. These conditions are often exacerbated by natural sources that can degrade air quality; for example, schools in two Indonesian cities were closed in 2019 due to smoke from forest and peat fires.

A host of urban air pollutants are known to put human health at risk. Fine and small particulate matter ($PM_{2.5}$ and PM_{10}) including sulfates, nitrates and black carbon, ground level ozone, and a number of additional inorganic and organic compounds can harm human health. Emissions from vehicles contribute to levels of these common pollutants and are also major sources of GHG emissions and toxins like benzene that are known to cause cancer.

The World Health Organization estimated that poor ambient air quality impacted 99% of the global population in 2019, resulting in 4 million premature deaths, with residents in poorer cities most affected. Examining data collected from 795 cities in 67 nations, the WHO reported that levels of PM in urban air samples increased by 8% in only 5 years, with the most troubling trends occurring in the eastern Mediterranean and Southeast Asia. Projections from the European Environmental Agency (Fig. 9.2) indicate there will be a corresponding increase in premature deaths in the coming decades.

In addition to elevating levels of criteria air pollutants, transportation also has a large carbon footprint, often because

J. Pontius, A. McIntosh, *Environmental Problem Solving in an Age of Climate Change*, Springer Textbooks in Earth Sciences, Geography and Environment, https://doi.org/10.1007/978-3-031-48762-0_9

Fig. 9.1 A haze of pollution hangs over Los Angeles. (Source: Diliff (CC BY SA 3.0) via Wikimedia Commons)

Fig. 9.2 Projected premature deaths due to particulate matter and ground-level ozone. (Source: European Environmental Agency {Public Domain})

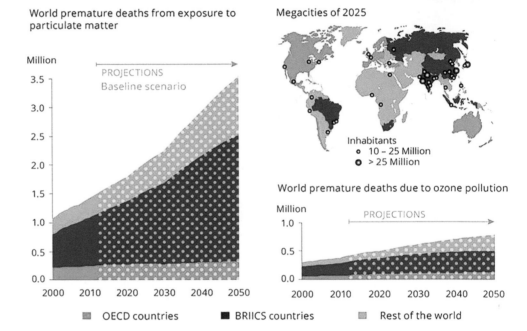

poor planning has led to vehicle-jammed streets and high-ways in sprawling urban and suburban areas with little alternative transport available. Cities like Los Angeles with its massive freeway system were designed around the automobile to meet the needs of populations that were once much smaller. The US EPA estimates that, currently, transportation is the largest and fastest-growing contributor to US GHG emissions.

9.2.2 The Role of Climate Change

Data from the US EPA (2018) underscore how much gasoline and diesel-powered vehicles contribute to climate change. One gallon of gasoline burned releases nearly 8900 g of CO_2, while the figure for a gallon of diesel is nearly 10,200 g. Driving the average passenger car 1.6 km releases 404 g of CO_2, and the typical passenger vehicle, driven on average about 18,500 km annually, emits about 4.6 metric tons of CO_2 per year. To understand the magnitude of impact on both climate change and air quality, consider that in 2021, there were 284 million cars on the road in the United States, with 1.5 billion worldwide.

While vehicle emissions contribute significantly to climate change, climate change also impacts air quality in many ways. Some of these connections between warming temperatures and air quality are obvious. For example, the chemical reactions that lead to the formation of pollutants like ozone are directly related to warmer temperatures (Fig. 9.3). But other connections may be more subtle. For example, warmer temperatures lead to longer growing seasons and an increase in airborne allergen pollutants. Below are lists of some of the direct and indirect ways that climate change is impacting air quality.

Direct effects:

- Warmer temperatures speed up the rates of chemical reactions, leading to higher levels of air pollutants like ozone.
- Less rainfall may lead to drought, drier soils, and more airborne particulate matter like wind-blown dust.
- Warmer temperatures and more rainfall may lead to increased growth of pollen-generating species, aggravating conditions for those suffering from allergies.
- Warmer temperatures may promote plant growth and greater formation of biogenic VOCs which also contribute to ozone formation.
- Worsening drought conditions may lead to more forest fires, threatening air quality in cities that lie downwind.
- Changing wind patterns and increased numbers of stagnation events may reduce atmospheric dilution of air pollutants in urban areas.
- Reduced precipitation can decrease the effectiveness of air pollutant removal.

Indirect effects:

- Increased air conditioner use in hotter cities will require more electricity, possibly generated by burning fossil fuels which release GHGs.
- Increased movement of environmental migrants into cities may lead to greater release of air pollutants from a variety of sources.

- Increased urban flooding from more intense storms and hurricanes can increase the formation of mold and other Indoor Air Pollutants.

In addition to the direct and indirect effects that climate change may have on urban air quality, it is important to recognize that there is also a positive feedback mechanism at work. As outlined above, climate change leads to increased air pollution in many ways. But increased air pollution can also contribute to warming. GHG emissions are, in and of themselves, pollutants that directly impact the climate.

But indirect feedbacks also exist, including positive feedback loops between pollutant production and a warming climate. For example, particulate matter (PM) pollution can travel long distances in the atmosphere, reaching polar regions where it darkens snow and ice and increases the absorption of solar radiation. This process accelerates warming at the poles, disrupting the jet stream, resulting in increased frequency and duration of droughts. This can lead to more wildfires, increasing PM emissions to the atmosphere and creating a reinforcing positive feedback cycle (Fig. 9.4).

9.2.3 Solutions

Because transportation is one of the largest sources of GHGs, ozone, hazardous pollutants, and VOCs, many solutions focus on ways to reduce emissions from gasoline and diesel-powered internal combustion engines. With growing concerns about air quality and human health, particularly in

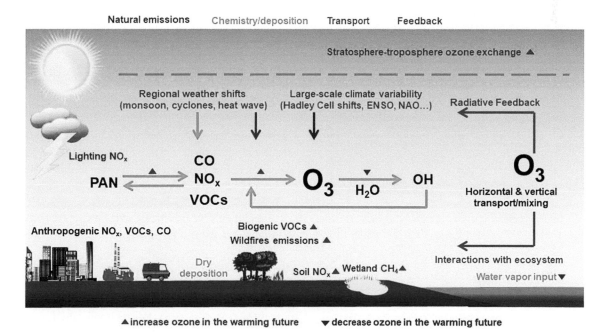

Fig. 9.3 Pathways of interaction between meteorology/climate change and tropospheric ozone. (Source: Lu et al. 2019 {Open Access})

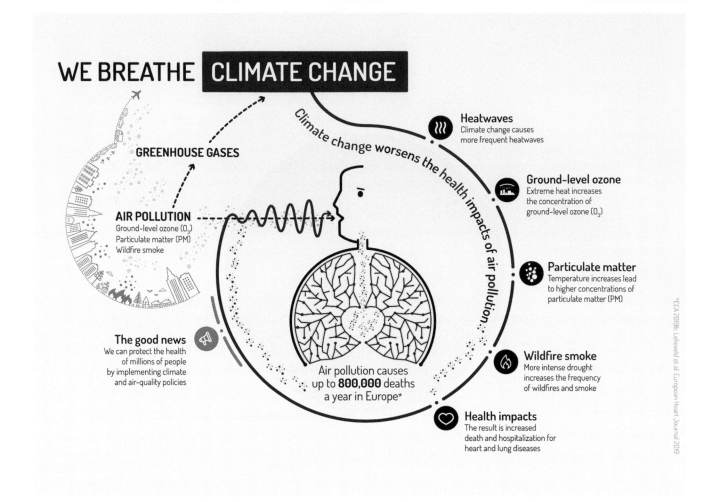

Fig. 9.4 The link between climate change, air pollution, wildfires, and human health. (Source: Exhaustion.eu. Illustration Info Design lab)

Fig. 9.5 Sign warning drivers that they are about to enter the Ultra-Low Emission Zone and Congestion Charging Zone. (Source: David Hawgood (CC BYSA 2.0) via Wikimedia Commons)

urban areas, progress has been made in identifying a number of possible steps to tackle this issue:

Financial Disincentives Some cities, like London (UK), are using financial disincentives to encourage drivers to switch to low-emission vehicles. City managers have established an Ultra-Low Emission Zone (ULEZ) (Fig. 9.5) in the center of the city where vehicles either meet a strict emission standard or pay both a daily usage fee and a daily Congestion Charge fee for driving in the most congested portions of the city during workday hours.

Carpooling Incentives Many urban areas are developing incentives to reduce single-occupancy vehicle (SOV) use. These range from free parking in cities to the construction of exclusive "car-pool" lanes on freeways. Despite such efforts, the majority of vehicles in most US cities are SOVs. San Jose CA found that SOV use declined only slightly below 80% in recent years despite such efforts.

Mass Transit From city buses to monorails, cities are increasingly focused on developing mass transportation options that get people out of their cars, reducing environmental impacts (Fig. 9.6). For example, the Las Vegas, NV monorail is estimated to have eliminated over 25 million vehicle miles since it began operation in 1995.

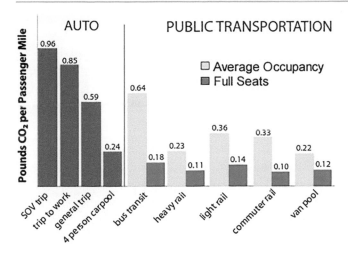

Fig. 9.6 Estimated CO_2 emissions per passenger mile for transit and private autos. (Source: US DOT Federal Transit Administration {Public Domain})

Alternative Fuel Vehicles There has been an explosion of new technologies for propelling motor vehicles. Electric, plug-in hybrid electric, and hybrid vehicles are becoming more popular, particularly with the success of companies like Tesla and Rivian. Tesla estimates that its electric vehicles had eliminated 5 million metric tons of CO_2 emissions by 2020. But while alternative fuel vehicles powered by ethanol, biodiesel, natural gas, and fuel cells may be cleaner options than gas and diesel, they come with their own set of environmental issues.

Urban Planning There is a growing consensus that better urban planning is critical for mitigating climate change. Forward-looking cities are building "smart" communities that place homes, businesses, schools, and commercial facilities in close proximity to minimize the need for automotive travel. This stands in contrast to US urban development patterns that necessitate the use of vehicles. Smart planning that includes a range of transportation choices and walkable neighborhoods can help improve air quality and reduce GHG emissions. Long-existing examples like BEDZED in the United Kingdom show that such approaches can work (Fig. 9.7).

9.2.4 Unit Challenge

The city of Cleveland, Ohio, is concerned about its air quality. The American Lung Association's 2021 "State of the Air" report noted that the city received a "failing" grade for ozone pollution and was the 14th most polluted US city for year-round levels of particulate matter (PM). The city wants to focus on reducing emissions from transportation in the city. While they are interested in long-term solutions like expanded mass transit, they also want to take short-term measures to reduce the amount of air pollutants released by urban traffic for more immediate relief.

You're in charge of the city's motor pool, which includes 200 gasoline-powered automobiles, 50 city buses that run on diesel, and 100 diesel-powered heavy-duty trucks. **Your Unit Challenge is to identify the best way to reduce air pollutants, including CO_2, released by the city's fleet that also considers environmental, social, and economic perspectives.** While your focus will be on reducing GHG emissions, you should also consider other air quality issues important to the city. You will make your recommendation in a one-page Fact Sheet to be presented to the Cleveland City Council.

9.2.5 The Scenario

The city has tasked you with developing a strategic plan to minimize emissions from the city's transportation fleet. Any solutions you propose must also maintain operational efficiency, minimize cost to taxpayers, and maximize community buy-in. After reviewing all the possible options to improve the fleet, you decide to focus on three:

1. **Hydrogen fuel cells.** New technology allows vehicles to run on hydrogen and emit only water vapor and oxygen as waste products. Such vehicles are more efficient than conventional internal combustion engine vehicles and produce no harmful tailpipe emissions. While old concerns about the safety and availability of hydrogen fuel have largely been answered, at least in your area of Ohio, there are still concerns about how the hydrogen is produced, with a host of "hydrogen colors" available, depending on the production process used.

   Specific recommendation: Replace half the gasoline-powered passenger vehicle fleet with hydrogen fuel cell models. There are several hydrogen fuel stations in the area, although the hydrogen available in the city is "gray" hydrogen, generated from natural gas.

2. **Compressed natural gas (CNG).** CNG, which is mostly propane with small amounts of butane and propylene, has a variety of uses, including fueling heating appliances and cooking equipment and powering vehicles. Readily attainable, even in rural areas, CNG is considered to be a cleaner burning fuel than gasoline and releases fewer particulates than burning diesel. It is, however, still a fossil fuel. While some have expressed concerns about the safety of CNG vehicles, there is no evidence to suggest that they are more dangerous than conventional fuel models.

   Specific recommendation: Convert the bus fleet from diesel to CNG. There are many such buses already on the road in many other cities, but safety concerns have been raised by a small group of citizens.

Fig. 9.7 The BedZED estate. BedZED stands for Beddington Zero Energy Development and is the United Kingdom's largest and first carbon-neutral eco-community. (Source: Tom Chance (CC BY 2.0) via Wikimedia Commons)

3. **Electric vehicles.** While most of the attention has been focused on the development of a market for electric (EV) passenger cars, EV commercial trucks have also taken hold. Orange EV, perhaps the largest commercial EV truck manufacturer in the United States, is in its 10th year of operation, serving 100 fleets across 24 states, Canada, and the Caribbean and exceeding one million hours and four million miles of operation.

   Specific recommendation: Convert the city's trucks to electric vehicles (EVs). With ample quick charging stations available in the Cleveland area, truck down time would be minimal. The question for EVs, however, is "where does the electricity come from?"

Each of these three approaches has important advantages and disadvantages and costs and benefits. Your task is to evaluate each and recommend the approach that will, in your opinion, provide the greatest reduction in transportation emissions, while maximizing stakeholder buy-in and minimizing impacts on the local community and economy.

9.2.6　Relevant Facts and Assumptions

- While each of the three alternatives you're considering releases, during its production and/or use, both GHGs and traditional pollutants like PM and NOx, you decide to focus on CO_2 released from the production and use of the alternative energy source as the basis for your recommendations.

- Assume that adequate funds will be available to implement any one of these solutions. However, operational costs will differ over the lifetime of the vehicles:

  - *Hydrogen Fuel Cell Cars*: Each of the 200 passenger cars in your fleet travels an average of 11,500 miles annually and averages 30 mpg on traditional gasoline that costs \$4/gallon. Hydrogen fuel costs about \$0.15/mi for these cars. Your plan would be to convert half of these cars to hydrogen fuel cell vehicles at an additional cost of \$5000 per vehicle. Converting from gas vehicles to hydrogen fuel cell vehicles is expected to reduce emissions by 14 kgCO_2/100 miles driven.

  - *Compressed Natural Gas Buses:* Each of the 50 buses travels an average of 25,000 miles annually and averages 17 mpg on traditional diesel fuel that costs \$4.50 per gallon. Natural gas fuel costs about \$0.29/mi for these buses. Your plan would convert all of the buses to CNG at an additional cost of \$15,000 per vehicle. Converting buses from diesel to CNG is expected to reduce emissions by 30 kg CO_2/100 miles driven.

  - *Electric Trucks*: Each of the 100 city trucks travels an average of 20,000 miles annually with an average of 20 mpg on traditional diesel fuel that costs \$4.50 per gallon. EV trucks travel about 2 mi/kWh at a cost of \$0.45/kWh. Your plan would convert all of the trucks in your fleet to EVs at an additional cost of \$7500 per vehicle. Converting from diesel trucks to electric trucks is expected to reduce emissions by 10 kg CO_2/100 miles driven.

9.2.7 Build Your Foundational Knowledge

Below are web sources that provide additional information about each of the solutions you're considering for this Unit Challenge. This information provides a critical foundation to help you evaluate each option and support your final choice. After reviewing each source, be prepared to answer questions in the Preparation Assessment Quiz and to summarize any information relevant to your Unit Challenge.

Hydrogen Fuel Cells:
Green Vehicle Guide: Hydrogen Fuel Cell Vehicles
Fuel Cell Vehicle Emissions

Compressed Natural Gas (CNG):
Natural Gas Vehicles: Why aren't we buying them?
Natural Gas Buses, Separating Myth from Fact

Electric (EVs):
Electric vehicle myths
How electric vehicles help tackle climate change

Final Product: A one-page Fact Sheet summarizing the issue, detailing your solution, and justifying your choice of that solution. Consider your audience, Cleveland's citizens and its city officials, and be sure to demonstrate how your proposed solution will stand up to the challenges posed by climate change.

9.2.8 Preparation Assessment Quiz

Are you ready to tackle your challenge? At this point you should understand the basic environmental principles and ecological processes involved in this environmental problem.

Consider the following questions. If you are comfortable with answering these, then you are ready to head into Discovery, Analysis, and Solutions activities.

- What are the six criteria air pollutants threatening urban areas?
- What are some of the ways that urban air pollutants affect human health?
- List two ways that climate change is likely to worsen urban air quality.
- Give an example of a disincentive cities can use to discourage automotive pollution.
- How can smarter urban planning help to improve air quality in cities?
- According to the US EOS's website on hydrogen fuel cell vehicles (FCVs), how are FCVs like EVs? How are they different?

- According to the Clean Cities Fact Sheet, what are two of the myths about CNG buses?
- Refute Myth #5 on the US EPA's web site about EVs and the climate.
- For each of the proposed solutions, are there any additional benefits that might arise from their implementation that might not be directly related to your Unit Challenge?
- For each of the proposed solutions, are there any negative unintended consequences that might result from their implementation?
- What additional information did you glean from your web sources that might help inform your Unit Challenge?

9.3 Urban Air Quality: Discovery

Specific Skills You'll Need to Review: Navigating the Scientific Literature, Science Communication, Problem-Solving

9.3.1 Independent Research

(Key Skill: Navigating the Scientific Literature)

To better understand the various approaches that could help reduce the amount of harmful emissions produced by Cleveland's motor pool vehicles, you first need to examine the literature to see what others have found. Conduct a search of the peer-reviewed scientific literature focused on the solution you have been assigned and identify one research paper that focuses on your assigned approach.

Prepare a summary of the article you selected that includes the following:

- **Citation**
- **Main topic**: Stick to a few words, likely pulled from the title.
- **General summary**: A few bulleted sentences summarizing the research question it addresses and approaches it takes.
- **Methods:** How did they approach their research question?
- **Location**: Where was the work done?
- **Conclusions**: Concise list of the findings, specifically capturing the take-home message.
- **Relevance**: How might this study help inform your Unit Challenge? Feel free to make a bulleted list of information you may want to include later.

9.3.2   Literature Share: Reciprocal Instruction

(Key Skill: Scientific Communication)

Share   In small groups, share and critique the research article you found. Keep in mind that your peers have not read this article, and it is your job to convey the key information to them. Note the items that will be important to consider when you are developing your solution to the challenge.

Critique   Evaluate how these studies might help inform your Unit Challenge. Consider the following:

- Source (Quality of the work or bias of the authors)
- Methods (Did their methods sufficiently address the research question?)
- Conclusions (Did the results justify the conclusions made?)
- Relevance (Can these findings be applied to your challenge?)

Based on your critiques, choose one article to share with the larger class, along with the key information that may be useful in deciding on a solution to propose.

9.3.3   Think-Compare-Share

(Key Skill: Problem-Solving)

Now that you have more information about possible solutions for this unit's challenge, you need to **develop a more formal problem definition** to guide your work throughout the rest of the exercises.

Think   Start by working independently to develop a specific Problem Statement to guide the remainder of your work. Problem Statements provide the relevant information and boundaries to make the issue something you can effectively assess and tackle. The basics of a formal Problem Statement include the following:

Problem Statement   A short, concise statement summarizing the issue that includes the following:

- A **description** of the undesired condition or change that you hope to achieve (What is the actual problem?)
- **Justification** for addressing the problem (Why does this problem matter?)
- Potential **sources** or **causes** of this problem (What is the cause you need to address?)

- The **metrics** you will use to assess the status of the problem (How will you know if you are making a difference in the problem?)
- The **desired outcome** for these metrics (What is the end goal or ideal state?)
- Potential **solutions** to consider (How might you attempt to achieve this goal?)

Compare/Share   Now return to your small group to share your Problem Statements. Use each of your ideas to develop a joint Problem Statement that contains all key information and is concise, clear, and well written.

Unit Discovery Summary
Submit a final Problem Statement that succinctly captures the key information to guide your work on this Unit Challenge.

9.3.4   Reflecting on Your Work

(Key Skill: Personal Reflection)

After your work in Discovery, you should have a better idea of the problems you face and have produced a Problem Statement you could use to tackle the Unit Challenge. Take a moment to reflect on this work. Consider the following prompts but feel free to expand on any to best capture your learning experience and better inform your next steps.

- Of the skills you practiced in Discovery, which were the most challenging? Which were the most interesting?
- How were you most comfortable working during these exercises? In small groups, independently, or with the larger class? Why? How does your choice reflect your personality type and leadership style?
- Was your Problem Statement strictly focused on the environmental problem of urban air quality, or did it also consider important social and economic considerations? How might a focus on the environmental aspects limit your ability to identify truly sustainable solutions?
- You've been given three viable solutions to assess as a part of this case study. But this is not an exhaustive list of options or even necessarily the best possible course of action for every scenario. Take a moment to "think outside the box." Are there any other possible solutions you think would be worth exploring? Describe one that you think would be worth pursuing.

9.4 Urban Air Quality: Analysis

Specific Skills You'll Need to Review: Quantitative Literacy, Sustainability Science

Review your Background and Discovery sections before beginning the Rotating Station exercises below. While you focused on one potential solution in your Independent Research in Discovery, keep an open mind as your work through Analysis activities.

9.4.1 Rotating Stations

(Key Skill: Quantitative Literacy)

At each of the following stations, you will review data that are relevant to the three potential solutions you're considering. Spend some time working through the analyses at each station to learn more about this issue and possible solutions for your Unit Challenge.

Be sure to write down one finding at each station that will help inform your selection of a solution.

Station 1: Hydrogen Fuel Cells Sources for the hydrogen required to operate a hydrogen fuel cell are varied. "Gray" hydrogen is produced using other fossil fuels like methane. "Blue" hydrogen adds carbon capture to the process, while "green" hydrogen is derived from renewable energy sources. Argonne National Laboratory's (ANL) report Fuel Choices for Fuel Cell Vehicles: Well-to-Wheels Energy and Emission Impacts compared GHG emissions for 10 of the most common hydrogen production and distribution pathways relative to emissions from traditional gasoline vehicles.

Based on the data in Fig. 9.8, answer the following questions:

- Are hydrogen fuel cells always a better alternative to gasoline engines? Explain why this might or might not be.
- Assume that if the 95% confidence intervals do not overlap, differences between hydrogen fuel sources are likely statistically significant. While there are obvious differences among the various energy sources, particularly among the gray (US energy mix sourced), blue (natural gas sourced), and green (renewable and solar sources) methods of H_2 production (as indicated by the colored circles), which ones are significantly better at reducing GHG emissions than the others?
- Based on these results, do you think we should be focused on liquid (LH_2) or gaseous (GH_2) production? Does it

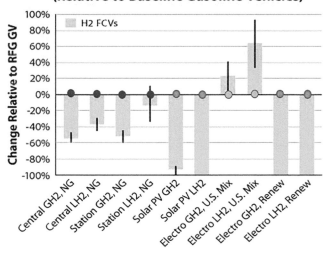

Fig. 9.8 Percent change in GHG emissions (with 95% confidence intervals) for all hydrogen fuel pathways relative to gasoline vehicles (zero values) including natural gas sources (NG), solar, a traditional US mix, and renewable sources of Gaseous Hydrogen (GH2) and Liquid Hydrogen (LH2) production that require transport (Station) as well as locally derived fuel (Central). "Electro" refers to hydrogen generated using electricity. (Source: Modified from US DOE Alternative Fuels Data Center {Public Domain})

depend on the hydrogen fuel source (denoted by colored dots)?
- If a 40% reduction in GHG emissions is your target, and you only have access to gray and blue sources of hydrogen fuel, which options allow you to meet your target?

Record any relevant Station 1 findings

Station 2: Compressed Natural Gas
Yuan et al. (2018) compared GHG emissions from traditional diesel, compressed natural gas, and liquefied natural gas across several types of vehicles in China. They considered all five stages of the fuel life cycle, from energy extraction through fuel use (C1, C2, C3, C4, and C5 in the figure below represent the following life cycle substages: energy extraction, energy transport, fuel production, fuel transport, and fuel use).

Based on the data presented in Fig. 9.9, answer the following questions:

- The data are presented as a per vehicle sum of GHG emissions. While all three bus fuel sources emit considerably more GHGs than passenger vehicles, consider that a bus typically carries 40 people, while a passenger vehicle typically carries 2. Estimate the total life cycle GHG emissions per person for the gasoline and CNG PVs compared to the per person diesel, CNG, and LNG bus options.

Based on this new metric, how do the buses compare to PVs in terms of total GHG emissions per person?

- How many passengers would a gas PV need to carry in order to match the per person emissions of a diesel bus? Conversely, how low would diesel bus ridership occupancy have to be to justify the use of natural gas-powered PVs?

- Until awareness and acceptance grow, initially adding new bus routes would likely attract only 20 riders per route. Would you recommend incentives to increase natural gas-powered PVs or use the incentives for the new diesel bus routes?

- Considering that there is little difference in GHG emissions between the diesel and natural gas-fueled buses, what other reasons might you give for conversion to natural gas besides GHG savings?

Record any relevant Station 2 findings

Station 3: Electric Vehicles

One of the challenges faced by the electric vehicle (EV) industry is social acceptance, with a number of concerns, including limited vehicle range per charge, long charging times, and poorly developed charging infrastructure. Burkert et al. (2021) studied social acceptance of EVs by the German public. The figure below shows changes in public opinion about a number of EV issues between 2011 (700 interviewees) and 2020 (1000 interviewees). Each point represents the percent of respondents concerned about the issues presented in each sector of the figure.

Based on the data presented in Fig. 9.10, answer the following questions:

- Considering all six issues in 2011 and 2020, how did the average percent of people concerned about various EV issues change? Consider the inverse of this metric as the percent who support the use of EVs. How many people supported the use of EVs in 2020?

- Which two of the six issues associated with use of EVs were of greatest concern to the Germans surveyed in 2011? In 2020?

- What do you think accounted for these changes in perception?

- Which of the six concerns do you believe is the most important to address in order to increase acceptance of EVs? Which do you believe would be the hardest to address? Why?

Record any relevant Station 3 findings

Station 4

An important consideration as you evaluate your three options is the cost of operating each type of vehicle. Given the information presented in Relevant Facts and Assumptions in the Unit Introduction, answer the following questions:

- Calculate the annual fuel costs for the current fleet, as well as annual fuel costs expected for each conversion solution (half of the passenger vehicles to hydrogen; all buses to CNG; all trucks to EVs)

- Now consider the additional cost per vehicle to convert your fleet under each proposed solution. Calculate the 10-year additional cost (above current annual operating costs) based on the annual fuel costs and total additional vehicle costs for each solution (note that additional vehicle costs are a one-time expense). Which solution is the least expensive?

- How many kg of CO_2 emissions could be saved per year with each of the three proposed solutions? Which solution has the potential to reduce CO_2 emissions the most?

- Based on your analysis, which solution is the most cost effective in terms of the conversion costs over 10 years relative to the emissions reductions achieved?

- How important should these costs be in the selection of any one potential solution?

Record any relevant Station 4 findings

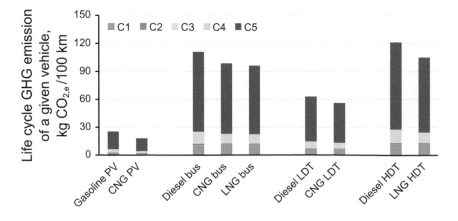

Fig. 9.9 Life cycle GHG emissions of Natural Gas Vehicles (NGVs) and comparable traditional fuel vehicles in China (PV=passenger vehicles; CNG=compressed natural gas; LNG=liquefied natural gas; LDT=light duty trucks; HDT=heavy duty trucks). (Source: Yuan et al. 2018 (Open Access))

Fig. 9.10 Comparison of social concerns about electric vehicle use between 2011 and 2020. (Source: Burkert et al. 2021 (Open Access))

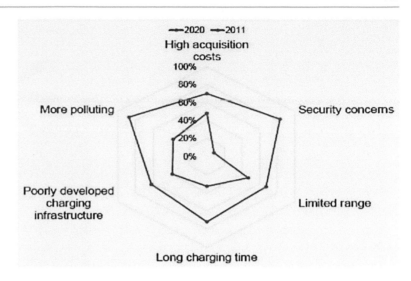

9.4.2 System Mapping

(Key Skill: Sustainability Science)

Your assigned solution may have a variety of direct and indirect economic, social, and ecological impacts that should also be considered. For instance, can the motor pool operate efficiently, given the additional time it may take to recharge your EVs? What about the social acceptability of "gray" hydrogen and electricity generated by burning coal?

Working with other class members assigned the same solution, develop a simple sustainability map that shows the various system connections across each of the three sustainability domains (ecological, economic, and societal). When you find connections between impacts that cross domains, draw a line to identify the connection (e.g., cheaper public transportation on CNG buses (economic) leading to increased financial resources for a family (social)).

The goal is to think broadly about this larger system and envision how implementing actions in one domain (e.g., ecological) may impact components in another domain (e.g., economic or societal).

Use the following template (Fig. 9.11) to get started:

When each group has completed their basic sustainability map, **come together as a class to compare maps for the three possible solutions.** This information will help inform your decision support work later in the unit.

Unit Analysis Summary *Based on your explorations, what have you learned that can help inform your choice of a solution? Do the data support the adoption of one or more than one of these potential solutions?*

9.4.3 Reflecting on Your Work

(Key Skill: Personal Reflection)

During your work in Analysis, you explored some of the research into possible solutions to help inform your decision. Take a moment to reflect on this work. Consider the following prompts but feel free to expand on any to best capture your learning experience and better inform your next steps.

- How did you feel working with data? Do you consider quantitative literacy a strength or an area for improvement for you?
- How important should science be in informing management and policy? Do you feel the data you examined support and justify the costs of addressing your challenge?
- Any solution should be examined using a sustainability science lens. Your goal is to improve Cleveland's air quality. But any solution you implement may have other direct and indirect impacts. What are other possible economic, social, or ecological impacts? How should these considerations influence your decision?

9.5 Urban Air Quality: Solutions

Review your Unit Challenge and the major findings from Discovery and Analysis, including the sustainability map you created to highlight connections among possible solutions and the larger socioecological system.

Specific Skills You'll Need to Review: Problem-Solving, Decision Support, Communicating Science

Fig. 9.11 Template sustainability map to help connect ecological, societal, and economic considerations associated with your potential solutions

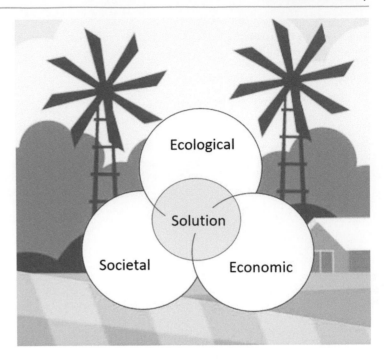

9.5.1 Small Group Guided Worksheets

(Key Skill: Decision Support)

Decision support matrices can help break down the desired outcomes to reflect multiple criteria for consideration, and they allow you to compare how each possible solution achieves those desired outcomes. This not only helps inform decision making, but it also provides transparency in the decision process and justification to help you advocate for adopting a particular solution.

For the three possible solutions, you will evaluate how well each can achieve the following desired outcomes:

- Minimizes CO_2 emissions from Cleveland's motor pool
- Minimizes costs
- Is a reliable way to meet the needs of the motor pool
- Is acceptable to motor pool drivers and community members
- Is a long-term solution (not a short-term fix), given a changing climate
- Minimizes secondary impacts (e.g., other air pollutants, human health and safety, local ecosystems, and economies)

Considering your three potential solutions, develop a formal decision support matrix to compare and evaluate each approach using the template matrix below (Table 9.1). Getting the most out of the decision matrix requires a depth of knowledge about each of the three possible solutions. Below we list additional sources about each option. Please

review each of these, paying particular attention to the two alternatives to your assigned approach.

Note that your group may have uncovered a novel solution not included in this list of three. You may choose to work through the structured decision matrix with your self-identified solution as a fourth solution option.

9.5.2 Additional Sources

1. Hydrogen fuel cells: Hydrogen fuel cells, explained | Airbus
2. Compressed natural gas: Alternative Fuels Data Center: Natural Gas Fuel Basics(energy.gov)
3. Electric (Evs): Study: Electric Vehicles Can Dramatically Reduce Carbon Pollution from Transportation, and Improve Air Quality | NRDC

Score each of your three potential solutions for each of the desired outcomes using a simple relative scale of 1 for least benefit to 5 for greatest benefit. Using a relative scale means you don't need to know exactly how well each solution meets the goal of each desired outcome, but you can use your judgment to assess how well each solution works compared to the others. *While this is a relative (subjective) scale, note that you will need to justify your scoring of each solution for each desired outcome.*

Once each cell in your decision matrix has a relative score, calculate an average score for each solution. Based on this analysis, which is the "best" solution, considering all your desired outcomes?

Table 9.1 Basic decision matrix to compare the three possible solutions for your Unit Challenge. Include your justification for each solution's ranking for each desired outcome

Solution Options	Desired Outcomes					
	Minimizes CO$_2$ emissions	Minimizes costs	Meets needs of motor pool	Provides a lasting impact	Maximizes community buy-in	Minimizes secondary impacts
Hydrogen fuel cell PVs						
CNG buses						
EV trucks						
Justification for Ranking						

Use a scale of 1-5 to score each option under each desired outcome category where 1 = does not achieve the desired outcome and 5 = completely achieves the desired outcome.

9.5.3 Role Playing

(Key Skill: Communicating Science)

In tackling environmental issues, you'll often find yourself working with groups of people with different perspectives about implementing a solution. You should use clear, concise communication to summarize your solution and justify its selection. To be effective, you must address concerns likely to be presented by various stakeholder groups and how the risks of taking these actions outweigh the risks of taking no action.

Your group will present and justify your chosen solution to the class with a particular emphasis on how it might benefit or impact one of the key stakeholder groups listed below. Be sure to include arguments that support this solution from the ecological, societal, and economic domains of sustainability science. Listeners will also be **assigned a stakeholder identity**, with an opportunity to **ask follow-up questions** after your presentation that reflect their unique concerns and perspectives. Your job will be to listen carefully and tailor your answers to this audience of stakeholders.

Key stakeholder groups include the following:

- Local residents who are most at risk from poor urban air quality
- Cleveland residents who rely on the vehicles in the motor pool
- Motor pool drivers
- Owners of companies selling gas- and diesel-powered vehicles
- State and federal air quality managers

9.5.4 Reflecting on Your Work

(Key Skill: Personal Reflection)

During your work in Solutions, you have explored several possible actions that could be taken to reduce CO$_2$ emissions in Cleveland. Take a little more time now to reflect on your findings and the skills you practiced. Consider the following prompts but feel free to expand on any of them to best capture your learning experience and feelings about this issue.

- Reflect on your work through a sustainability science lens. Does your solution address ecological, economic, and societal considerations? Which considerations do you think should carry the most weight in the decision? Why?
- How did you weigh solutions that might have the greatest environmental impact against those that are most likely to be implemented and maintained for long-term impact?
- How can environmental scientists work to show the value of healthy ecosystems and justify the costs of mitigation strategies?
- Science communication can be challenging, especially when working with diverse audiences. We need to craft our communication to match the interests and values of the target audience, but how do you do this when your audience contains a mix of stakeholder groups? How can you maximize the impact of your message to a diverse audience?

Unit Solution Summary *Summarize and justify your final solution choice and outline how it addresses the direct challenge while also considering social, economic, and ecologi-*

cal impacts. Also demonstrate that it will continue to meet the challenges posed by climate change.

9.6   Urban Air Quality: Final Challenge

As a part of this Unit Challenge, you were asked to write a one-page Fact Sheet justifying your choice for an approach to reduce GHG emissions from the city of Cleveland's vehicle fleet. Your Fact Sheet should include the following components:

- Brief problem statement
- Recommended mitigation strategy with sufficient details to summarize the general approach
- Justification of this recommendation (e.g., long-term effectiveness, given anticipated climate changes, implementation costs, other benefits provided, etc.). Be sure to use a sustainability lens to include considerations of direct and indirect ecological, social, and economic considerations
- Any obstacles the group might face trying to implement your solution

Consider the use of figures, graphics, and tables to help summarize the system and how this solution is well suited to meet all desired outcomes.

Final Unit Challenge Submit your final recommendations in a one-page Fact Sheet using clear science communication designed for a lay audience.

References

American Lung Association (2021) State of the air. https://lung.org

Burkert A, Fechtner H, Schmuelling B (2021) Interdisciplinary analysis of social acceptance regarding electric vehicles with a focus on charging infrastructure and driving range in Germany. World Electr Veh J 12:25. https://doi.org/10.3390/wevj12010025

Lu X, Zhang L, Shen L (2019) Meteorology and climate influences on tropospheric ozone: a review of natural sources, chemistry, and transport patterns. Curr Poll Rep 5:238–260

Massachusetts Department of Environmental Protection (2022) Transportation & air quality commonwealth of Massachusetts. https://www.Mass.gov/guides/transportation_air_quality

United Nations (2019) Cities: a 'cause of and solution to' climate change. UN News. Global perspective: Human stories. https://news.un.org/en/story/2019/09/1046662

U.S. Environmental Protection Agency (2018) Greenhouse gas emissions from a typical passenger vehicle. EPA-420-F-18-008. https://www.epa.gov/greenvehicles/greenhousegas-emissions-typical-passenger-vehicle

Yuan JH, Zhou S, Peng TD, Wang GH, Ou X-M (2018) Petroleum substitution, greenhouse gas emissions reduction and environmental benefits from the development of natural gas vehicles in China. Petrol Sci 15:644–656. https://doi.org/10.1007/s12182-018-0237-y

Core Knowledge
Urban ecosystems, Energy conservation, Sustainable development, Green infrastructure

10.1 Environmental Issue

Often offering better job opportunities and a refuge from social and political strife, urban areas are currently home to half of the global population. Experts believe that by 2050, that percentage will increase to two thirds. Unfortunately, this growth is typically opportunistic, with minimal consideration of environmental concerns. Sustainable development and green building practices are widely accepted concepts, but existing infrastructure and cost concerns often make it difficult for local officials to adopt such practices.

The heart of a city lies in its built environment: buildings, roadways, and other structures. These structures have substantial direct impacts on the environment, including effects on air and water quality, the distribution of natural habitats, and the richness of biological diversity. Recently, direct impacts of the urban built environment on local temperatures have been identified as a particular threat to human health. The urban heat island (UHI) effect is caused by a high density of surfaces like pavement and buildings that absorb and retain heat, increasing local temperatures to well above those of surrounding landscapes (Fig. 10.1). With climate change already affecting large cities, it is imperative to plan and develop greener built environments to reduce the threat posed by UHIs. In this unit, you'll explore how UHIs may be managed in a warming climate.

10.2 Background Information

10.2.1 The Problem

The design of many cities has led to an outsized environmental footprint and unique vulnerability to the challenges of climate change. While many city managers are now beginning to plan for a climate-disrupted future, their job is made more difficult because of the physical structure of the city already in place and the financial investment needed to undertake critical mitigation projects. While there are many environmental issues associated with how cities are designed and built, this unit will focus on the urban heat island (UHI) effect, with a look at how and why it forms, its impacts on urban dwellers, how climate change will exacerbate its effects, and some possible solutions.

Different surfaces absorb and reflect solar radiation in different ways. Vegetation absorbs certain wavelengths which drive photosynthesis while reflecting a large portion of near-infrared energy back from its surfaces. In contrast, most of the materials that constitute the built infrastructure, like buildings, roofing, and pavement, absorb a large percentage of solar radiation, which they then re-emit as thermal wavelengths (heat). This leads to significant additional heating in pockets or "island" areas which are largely covered by built surfaces.

While impervious surfaces play a major role in UHI formation, the absence of trees and other vegetation in built environments also plays a role. Not only do trees provide cooling shade, but they also help reduce ground-level temperatures via evapotranspiration. As the landscape is cleared for urban development, the cooling benefits provided by vegetation are lost.

J. Pontius, A. McIntosh, *Environmental Problem Solving in an Age of Climate Change*, Springer Textbooks in Earth Sciences, Geography and Environment, https://doi.org/10.1007/978-3-031-48762-0_10

Fig. 10.1 A thermal image of Atlanta, GA shows a *27 °C* air temperature with many surface temperatures exceeding 48 °C. (Source: NASA/Goddard Space Flight Center via Wikimedia Commons {Public Domain})

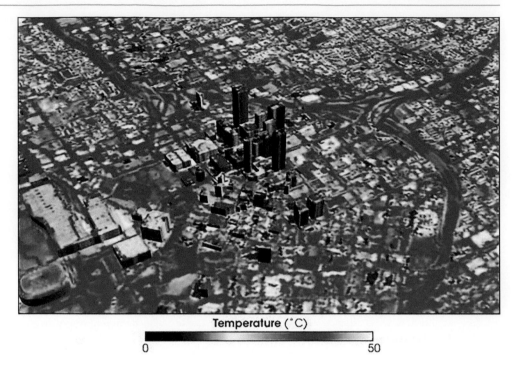

Temperature (˚C)

0   50

When the summer sun strikes exposed urban surfaces like pavement and roofs, the temperatures of these surfaces can increase 27 to 50ºC. On average, the difference in daytime surface air temperatures between urban and less-developed areas can be as much 15 °C (US EPA, 2008) (Fig. 10.2). The global nature of the problem was underscored by a report by Santamouris (2020) which indicated that the magnitude of urban temperature increases in global cities may exceed 4 to 5 °C and, at peak, may reach 10 °C. The authors estimated that by 2020, more than 400 cities worldwide were already experiencing urban overheating.

In addition to increasing urban temperatures, UHIs can lead to other impacts. Elevated air temperatures promote ozone formation and raise the risk of illnesses and death among urban residents. The US Centers for Disease Control and Prevention recorded an average of 702 heat-related deaths per year between 2004 and 2018. Older adults, young children, and people in poor health tend to be particularly vulnerable to the combination of extreme heat and impaired air quality. Low-income populations are also disproportionately affected due to poor housing conditions and limited access to medical resources.

High urban surface temperatures can also impact water quality, primarily by increasing the temperature of stormwater runoff that enters nearby waterways. Water temperature affects all aspects of aquatic ecosystems, with rapid changes in temperature being particularly stressful to aquatic life.

There are also significant economic and societal effects from UHI formation. The increased demand for electricity to cool buildings can stress the electrical infrastructure, possibly resulting in more frequent or prolonged power outages. If

temperatures were to increase 3.5 to 5.0 °C, such a change could increase the need for additional electrical generating capacity by roughly 10–20% by 2050. This would cost hundreds of billions of dollars. A recent report indicated that climate change is likely to be twice as costly in cities compared to rural areas.

On the flip side, in areas of extremely cold winters, UHIs can help mitigate the worst effects of the cold on human health and can reduce the demand for heating in buildings. Scientists are now working to develop new color-changing building materials that could reduce the UHI effect in the summer, while keeping colder cities warmer in the winter.

10.2.2 The Role of Climate Change

Climate change and urban heat islands are directly connected, with positive feedbacks working in both directions. The warming climate will have substantial impacts on the built environment and the development of UHIs. Higher temperatures associated with climate change are already leading to longer, more severe heatwaves (Fig. 10.3). Areas already affected by heat islands will likely bear the brunt of these heatwaves and their associated harmful health and environmental effects.

However, UHIs are also impacting climate change. The increased demand for cooling of residential buildings during heatwaves requires greater consumption of energy, with the US EPA (2008) noting that peak demand increases 1.5 to 2 percent for every 1°F increase in ambient temperature. This electrical demand is often met by power plants that burn fos-

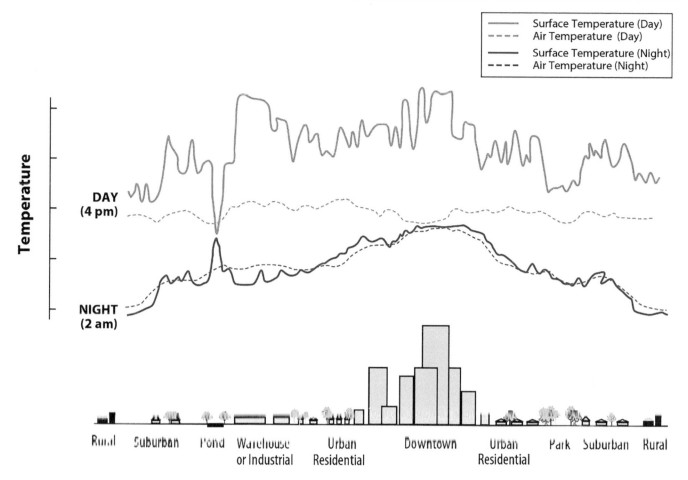

Fig. 10.2 Surface temperatures vary more than atmospheric air temperatures during the day, but they are generally similar at night. The dips and spikes in surface temperatures over the pond area show how water maintains a nearly constant temperature day and night because it does not absorb the sun's energy the same way as buildings and paved surfaces. Parks, open land, and bodies of water can create cooler areas within a city. Temperatures are typically lower at suburban-rural borders than in downtown areas. (Source: US EPA {Public Domain})

sil fuels. Such heavy demand can then result in substantial releases of pollutants, including GHGs, to the atmosphere. Increased GHG emissions hasten climate change, leading to more urban heatwaves.

Climate migrants will also impact the development and expansion of UHIs. To accommodate the influx of migrants expected to flood cities in the future, more natural landscapes will be lost. Not only do trees and other plants absorb CO_2, but they also provide shading to help cool urban dwellers. The conversion of vegetated surfaces to built structures as cities increase in size will only worsen the effects of climate change.

These important interactions indicate that continued warming will worsen UHIs, and the expansion of heat islands will lead to more GHG emissions, which, in turn, will exacerbate climate change. But communities can take steps to mitigate these trends by lowering UHI temperatures and reducing GHG emissions.

10.2.3 Solutions

While tackling UHIs will be challenging for many cities, given the limitations of existing infrastructure and considerable economic investment required to tackle the problem, there are many ways that cities can work to mitigate impacts. These include immediate measures to help protect vulnerable populations as well as a shift to green building and sustainable development planning to reduce UHI effects.

Protecting Human Health There are a number of examples of major cities already dealing with some of the effects of UHIs. For example, a 2010 heatwave killed 1300 in Ahmedabad, a city of 8 million in western India. In 2013, the city implemented a heat action plan, with steps such as opening public buildings and mosques as cooling centers and providing ice packs to outdoor workers. Deaths during a similar heatwave in 2015 totaled fewer than 20.

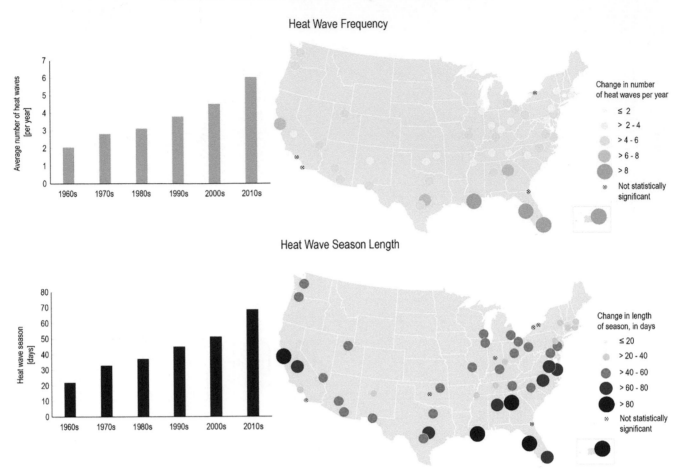

Fig. 10.3 Changes in the number of heatwaves per year (frequency) and the number of days between the first and last heatwave of the year (season length). These data were analyzed from 1961 to 2019 for 50 large US metropolitan areas. The graphs show averages across all 50 metropolitan areas by decade. The size/color of each circle in the maps indicates the rate of change per decade. Hatching represents cities where the trend is not statistically significant. (Source: Globalchange. gov {Public Domain})

Increasing Green Space Revegetating built environments can help reduce temperatures in UHIs. In some cities, replanting shade trees has been underway for decades. According to Shickman and Rogers (2020), to date, Sacramento Shade has planted over half a million trees throughout Sacramento County CA. Ko et al. (2015) analyzed 22 years of tree survival, tree growth, and energy savings related to the program. The authors found that Sacramento Shade had a 22-year post-planting tree survival rate of 42% and that annual energy savings from the program were 107 kWh per property and 80 kWh per tree.

Other cities are developing comprehensive revegetation plans to deal with projected climate conditions. For example, officials in Los Angeles have done a comprehensive study of the UHI in that city (Borunda, 2021). Poorer areas of the city have fewer trees to provide shade and suffer the most when high temperatures affect the city. In response, city officials have announced that they will plant 90,000 trees in the most heat-prone parts of the city. Baltimore, MD is hiring residents to plant trees in underrepresented neighborhoods, while Phoenix, AZ is targeting primarily poorer Hispanic areas for tree plantings and shade structures.

A team of researchers recently examined which types of structures and green spaces were the most effective at cooling the surrounding air on a city block. They found that green spaces with depth and multiple layers and heights of vegetation (including a mix of tall trees, shrubs, and ground cover) (Fig. 10.4) provided the most cooling, as opposed to green spaces comprised of a single line of tall trees (Park et al. 2017).

Reflective Surfaces in the Built Environment Often, fairly simple steps can make a big difference in large cities. Scientists and engineers have been working on ways to reflect sunlight before it can raise temperatures. Approaches

Fig. 10.4 Novel architectural designs allow for additional green space to be included on vertical structures as well as ground surfaces. (Source: CC0 via pxhere.com {Public Domain})

Include using new types of building supplies such as heat reflecting aggregates or surface coatings for cooler roofing materials and innovative pavement materials for parking lots and roadways. Recent research in California where cool pavements have been deployed showed that increasing reflectance of paved surfaces by 20% resulted in a reduction in outdoor temperatures of about 0.5 °C (Levinson et al. 2017). One interesting indirect effect of using these increased reflectance surfaces was a reduced need for streetlights to illuminate roadways.

Green Building Design Indoor environments will also be impacted as climate change worsens the effect of UHIs. Reducing energy consumption and carbon emissions associated with cooling buildings is one approach for mitigating the impacts of UHIs. According to the International Energy Agency (2018), air conditioning and electric fans used to cool buildings account for about 10% of global energy use, with GHG-emitting fossil fuels meeting much of this electrical demand.

Innovations in building materials like ultra-low U-value glazing for windows and seasonal energy storage provide a pathway for buildings with essentially zero heating demand. Efforts are also underway to minimize energy required to cool buildings through passive cooling design. Passive cooling uses insulating building materials, thermal mass, shade structures, window placement, and natural ventilation to reduce heat gain in buildings. Design features vary from dry arid to warm tropical climates, but they can be an effective way to maintain comfortable conditions without air conditioning.

Currently, there is a focus on implementing design strategies that also mitigate the contribution of buildings to surrounding temperatures. Green roofs or rooftop gardens can remove heat from the air as well as provide additional shade for the buildings they cover (Fig. 10.5). Green roof temperatures can be up to 20 °C cooler than traditional roofs and can reduce surrounding ambient temperatures by 2 to 3°C. In addition, green roofs can improve air quality by absorbing pollutants and can also mitigate stormwater discharge.

Climate change will likely impact built environments in a number of ways, and given the massive financial outlays that will be needed in many cities, coping with the effects of climate change on UHIs will require city managers to make difficult choices about which changes to implement first to minimize threats to human life and property.

10.2.4 Unit Challenge

Consider that you have been hired as an environmental planner to help a medium-sized city reduce its UHI effect and carbon footprint. Specifically, you are being asked to consider both changes to existing buildings, guidelines for new construction as the city grows, and steps to reduce outside ambi-

Fig. 10.5 Buildings on the World Intellectual Property Organization campus in Geneva, Switzerland, feature green roofs with more than 40 varieties of flowering plants and grasses to support biodiversity, improve rooftop thermal insulation, reduce ambient temperatures, and allow for better stormwater drainage. (Source: Emmanuel Berrod [CC BY 4.0] via Wikimedia Commons)

ent temperatures during the hottest months. Your plan must consider the long-term needs of the community as it continues to grow, as well as the impacts of ongoing climate change.

10.2.5 The Scenario

The city has hired you, an environmental planner with a strong background in green building and sustainable development, to help them identify a plan to both reduce ambient temperatures in their city and minimize GHG emissions from the built environment.

After reviewing the literature and the existing built environment of the city, you narrow down possible approaches to the following three focus areas:

1. **Planting shade trees**. There are many benefits to increasing tree cover and dispersal of urban greenways. Whether focused on planting street trees, building up vegetation in parks and gardens, or allowing natural revegetation of abandoned lots, additional vegetative cover in urban areas supports wildlife and biodiversity, improves mental and physical wellbeing of residents, enhances aesthetics, purifies stormwater runoff, and can dramatically reduce ambient air temperatures in UHIs.
   Specific recommendation: Convert 100 ha of publicly owned, currently unvegetated properties throughout the downtown area to fully vegetated conditions.
2. **Installing green roofs**. A green roof consists of vegetation planted over a waterproofing layer installed on top of a flat or slightly sloped roof. These roofs provide shade, reduce thermal loads on buildings, improve stormwater quality, and reduce ambient air temperatures. In addition, they provide an economic benefit. Over an estimated

lifespan of 40 years, a green roof can save about $100,000 in reduced energy costs compared to a 1000 m² conventional roof.
   Specific recommendation: Install green roofs on a quarter of the 500 buildings in the downtown area.
3. **Initiating a "Cool Roofs" program.** "Cool roofs" are constructed of materials that reflect sunlight and absorb less heat. Depending on the location, cool roofs reduce indoor temperatures by 2 to 5 °C compared to traditional roofs. Cool roofing materials can also enhance durability and improve the appearance of roofs. Government programs across the globe have successfully supported the transition from conventional to cool roofs, often with a focus on supporting lower-income areas to help mitigate the unequal burden that such populations often bear.
   Specific recommendation: Replace conventional roofing with cool roofing materials on half of the 500 buildings in the downtown area.

10.2.6 Relevant Facts and Assumptions

- The city of interest is located in a temperate climate with a population of about 200,000 permanent residents and a concentrated downtown area dominated by multistory businesses and apartment buildings.
- The average roof is 1000 m². Average annual cooling costs for each building are $5000.
- *Cool Roofs Initiative*: The goal is to install new reflective roof surfaces on half the 500 buildings in the downtown area. Installation costs about $12 per m² and can reduce thermal load (cooling costs) in the buildings by about 10% while reducing surrounding air temperatures by 1.8°C. The average lifespan is approximately 20 years.

- *Green Roofs:* The goal is to install green (vegetated) roofs on a quarter of the 500 buildings in the downtown area. Installation of green roofs costs about \$27 per m^2 and can reduce the thermal load in the buildings by about 25% while reducing surrounding air temperatures by 2.9°C. The average lifespan is approximately 30 years.
- *Urban Tree Canopy:* Across the 1000-ha city, 100 ha are appropriate for street tree plantings adjacent to buildings. Estimates for an urban tree canopy planting program are about \$5750 per ha using volunteer labor for planning and maintenance. Once canopies mature, shade from these trees could reduce cooling costs in nearby buildings (about one tenth of the total buildings in the city) by about 15%, while reducing air temperatures citywide by 0.3°C. With proper management, the lifespan of such green spaces is approximately 50 years.

Each of these three approaches has important advantages and disadvantages and costs and benefits. Your task is to evaluate each and recommend the one that will, in your opinion, provide the greatest reduction in both ambient air temperatures and GHG emissions at the lowest price, while maximizing additional benefits and minimizing impacts on the local economy.

10.2.7 Build Your Foundational Knowledge

Below are web sources that provide additional information about each of the solutions you're considering for this Unit Challenge. This information provides a critical foundation to help you evaluate each option and support your final choice. After reviewing each source, be prepared to answer questions in the Preparation Assessment Quiz and to summarize any information relevant to your Unit Challenge.

Shade Trees:
First of its kind study quantifies how tree shade can cancel urban heat island effect,
Using trees and vegetation to reduce heat islands

Green Roofs:
Green roofs
How green roofs and cool roofs can reduce energy use, address climate change and protect water resources in southern California

"Cool Roof" reflective surfaces:
Energy Saver: Cool Roofs,
Cool roofs fight climate change

Final Product: A one-page Fact Sheet summarizing the issue, detailing your solution, and justifying your choice of that solution. Consider your audience, town officials, and business owners, and be sure to demonstrate how your proposed solution will stand up to the challenges posed by climate change.

10.2.8 Preparation Assessment Quiz

Are you ready to tackle your challenge? At this point you should understand the basic environmental principles and ecological processes involved in this environmental problem. Consider the following questions. If you are comfortable with answering these, then you are ready to head into Discovery, Analysis, and Solutions activities.

- What is the UHI effect and what are its primary causes?
- What are the various ways that climate change and UHIs interact?
- How is UHI formation an environmental justice issue?
- What are some of the ways that city planners can mitigate the impact of UHIs?
- According to the US EPA's report on shade trees, what is the greatest contributor to the annual cost of maintaining urban forests?
- The NRDC's report on green and cool roofs estimates that if 50% of the roof surfaces in Southern California had green or cool roofs, how great would the energy savings be annually?
- According to the Energy Saver's report, most cool roofs have a high "thermal emittance." What is "thermal emittance"?
- For each of the proposed solutions in your Unit Challenge, are there any additional benefits that might arise from their implementation that might not be directly related to the challenge?
- For each of your proposed solutions, are there any negative unintended consequences that might result from their implementation?
- What additional information did you glean from your web sources that might help inform your Unit Challenge?

10.3 Urban Heat Islands: Discovery

Specific Skills You'll Need to Review: Navigating the Scientific Literature, Science Communication, Problem-Solving

10.3.1 Independent Research

(Key Skill: Navigating the Scientific Literature)

To better understand the various approaches being considered to reduce the UHI effect in your city, you first need to examine the literature to see what others have found. Conduct a search of the peer-reviewed scientific literature focused on the solution you have been assigned, and identify one research paper that focuses on your assigned approach.

Prepare a summary of the article you selected that includes the following:

- **Citation**
- **Main topic**: Stick to a few words, likely pulled from the title.
- **General summary**: A few bulleted sentences summarizing the research question it addresses and approaches it takes.
- **Methods:** How did they approach their research question?
- **Location**: Where was the work done?
- **Conclusions**: Concise list of the findings, specifically capturing the take-home message.
- **Relevance**: How might this study help inform your Unit Challenge? Feel free to make a bulleted list of information you may want to include later.

10.3.2 Literature Share: Reciprocal Instruction

(Key Skill: Scientific Communication)

Share In small groups, share and critique the research article you found. Keep in mind that your peers have not read this article, and it is your job to convey the key information to them. Note the items that will be important to consider when you are developing your solution to the challenge.

Critique Evaluate how these studies might help inform your Unit Challenge. Consider the following:

- Source (Quality of the work or bias of the authors)
- Methods (Did their methods sufficiently address the research question?)
- Conclusions (Did the results justify the conclusions made?)
- Relevance (Can these findings be applied to your challenge?)

Based on your critique, choose one article to share with the larger class, along with the key information that may be useful in deciding on a solution to propose.

10.3.3 Think-Compare-Share

(Key Skill: Problem Solving)

Now that you have more information about possible solutions for this unit's challenge, you need to **develop a more formal problem definition** to guide your work throughout the rest of the exercises.

Think Start by working independently to develop a specific Problem Statement to guide the remainder of your work. Problem Statements provide the relevant information and boundaries to make the issue something you can effectively assess and tackle. The basics of a formal Problem Statement include the following:

Problem Statement: A short, concise statement summarizing the issue that includes the following:

- A **description** of the undesired condition or change that you hope to achieve (What is the actual problem?)
- **Justification** for addressing the problem (Why does this problem matter?)
- Potential **sources** or **causes** of this problem (What is the cause you need to address?)
- The **metrics** you will use to assess the status of the problem (How will you know if you are making a difference in the problem?)
- The **desired outcome** for these metrics (What is the end goal or ideal state?)
- Potential **solutions** to consider (How might you attempt to achieve this goal?)

Compare/Share Now return to your small group to share your Problem Statements. Use each of your ideas to develop a joint Problem Statement that contains all key information and is concise, clear, and well written.

Unit Discovery Summary *Submit a final Problem Statement that succinctly captures the key information to guide your work on this Unit Challenge.*

10.3.4 Reflecting on Your Work

(Key Skill: Personal Reflection)

After your work in Discovery, you should have a better idea of the problems you face and have produced a Problem Statement you could use to tackle the Unit Challenge. Take a moment to reflect on this work. Consider the following prompts but feel free to expand on any to best capture your learning experience and better inform your next steps

- Of the skills you practiced in Discovery, which were the most challenging? Which were the most interesting?
- How were you most comfortable working during these exercises? In small groups, independently, or with the larger class? Why? How does your choice reflect your personality type and leadership style?
- Was your Problem Statement strictly focused on the environmental problem of UHIs, or did it also consider important social and economic considerations? How might a focus on the environmental aspects limit your ability to identify truly sustainable solutions?
- You've been given three viable solutions to assess as a part of this case study. But this is not an exhaustive list of options or even necessarily the best possible course of action for every scenario. Take a moment to "think outside the box." Are there any other possible solutions you think would be worth exploring? Describe one that you think would be worth pursuing.

10.4 Urban Heat Islands: Analysis

Specific Skills You'll Need to Review: Quantitative Literacy, Sustainability Science

Review your Background and Discovery sections before beginning the Rotating Station exercises below. While you focused on one potential solution in your Independent Research in Discovery, keep an open mind as your work through Analysis activities.

10.4.1 Rotating Stations

(Key Skill: Quantitative Literacy)

At each of the following stations, you will review data that are relevant to the three potential solutions you're considering. Spend some time working through the analyses at each station to learn more about this issue and possible solutions for your Unit Challenge.

Be sure to write down one finding at each station that will help inform your selection of a solution.

Station 1: Green Roofs Green roofs can reduce energy consumption and improve indoor comfort, particularly in hot climates. Wahba et al. (2018) modeled the impact of adding green roofs and walls to existing buildings in Cairo, Egypt. The figure below compares monthly cooling loads in kWh for buildings of various sizes with (A) no green features and (E) both a green roof and green walls.

Answer the following questions about energy savings associated with green building features in Fig. 10.6:

- How does the impact (reduced kWh required for cooling) of green design features (E) differ from conventional design (A) for buildings of different sizes (5, 10, and 15 floors)?
- How does the impact (reduced kWh required for cooling) of green design features (E) differ from conventional design (A) across months?
- Calculate the difference in kWh between A and E (without and with green features) for the high-load months (July and August) for large buildings (15 floors).
- If electrical energy in this region costs $0.15/kWh, how great could savings be for large buildings with green design features during these high-demand months?
- How might this savings differ for cities located in more temperate regions (e.g., what would you expect if the models had been run for London, United Kingdom, instead of Cairo, Egypt)?

Record any relevant Station 1 findings

Station 2: Cool Roofs Tewari et al. (2019) modeled the UHI effect for a heatwave in two urban environments (Phoenix, AZ, and New York City) (Fig. 10.7). They considered three green development scenarios where infrastructure was held constant, replaced with a green roof, or replaced with cool roof systems. Their goal was to see if the projected increase in the urban heat index expected with climate change could be offset by adopting irrigated green roofs or cool roofs in urban areas.

Based on the data in the figures above, answer the following:

- Describe the differences in the change in temperature (ΔT°C heat island index) between Phoenix (left figures) and New York City (right figures). What might account for these differences?
- What might account for the variability in ΔT °C over time in these figures?
- What was the impact of the simulated green and cool roof infrastructure in both cities? What does this tell you about the contribution of buildings to the UHI effect?
- Which type of treatment (Green roof or Cool roof) best minimized the UHI effect? Why might this be?

Record any relevant Station 2 findings

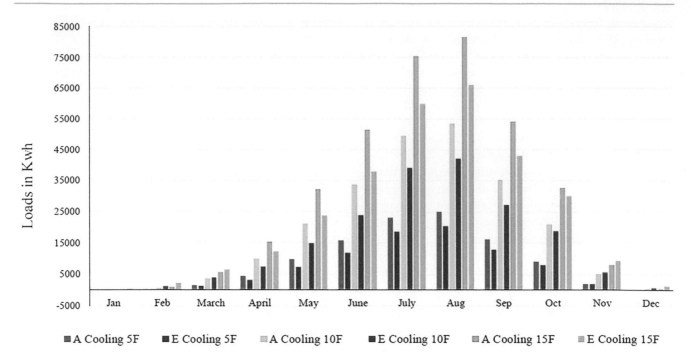

Fig. 10.6 A comparison of air-cooling annual loads between building options A (without green features) and E (with green features) for buildings with 5, 10, and 15 floors (F). (Source: Wahba et al. 2018 (Open Access-CC-BY 4.0))

Station 3: Urban Tree Canopy Scientists have demonstrated the important role that urban forests can play in reducing the effects of UHIs. Dialesandro et al. (2019) examined 10 different urban centers worldwide to look at the temperature differences among urban areas, urban forests, and areas adjacent to urban forests during the day and night.

Table 10.1 shows some interesting differences among metro areas, urban forest, and urban forest adjacent areas. Use this information to answer the following questions:

- Are urban forests always cooler than metro urban areas? Considering the cities where temperature differences might not be as large, why might this be?
- Are temperature differences greater during the day or at night? Why might this be?
- How do temperature changes measured in urban forests affect urban forest adjacent areas? What does this tell you about the ability of urban forests to impact temperatures more broadly throughout UHIs?
- This study included results from 10 arid cities. Consider that vegetation in more temperate regions can further reduce temperatures in urban forests and adjacent areas by an additional 2 °C. What would be the average temperature reduction in temperate urban forests in more temperature regions? In urban forest adjacent areas?

Record any relevant Station 3 findings

Station 4 For your Unit Challenge, you were given a set of assumptions to help guide your decisions. Now do some calculations to estimate how much each approach would cost to implement and how much in cooling costs (as a proxy for GHG emissions) each could be expected to save.

Refer to your Relevant Facts and Assumptions to calculate the following for each approach:

- How much would it cost to implement each of your solutions in the city?
- What are the total expected energy savings over the lifetime of each solution?
- Assume that the overall temperature reduction citywide is proportional to the percent of buildings converted under each method. Estimate the citywide reduction in temperature over the lifetime of each solution (sum temperature reduction over years).
- Considering that each solution has a different impact on citywide air temperatures, normalize the effective citywide reduction to a cost per degree cooling over the lifetime of each solution. Which provides the greatest cooling impact per dollar over its lifetime?
- How much should implementation costs weigh in your selection of a solution?

Record any relevant Station 4 findings

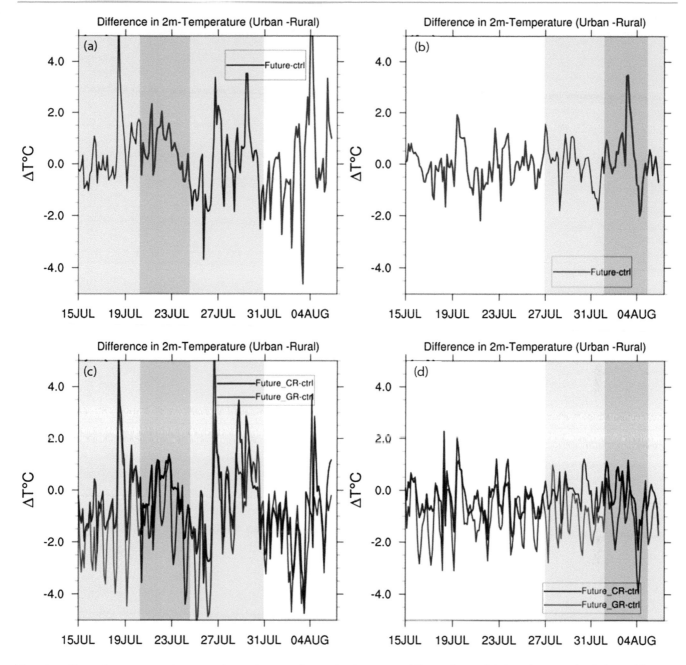

Fig. 10.7 Change in temperature between urban and surrounding rural landscapes in a simulated heatwave for Phoenix, AZ (left) and New York City (right). Note that panels (**a**) and (**b**) represent the future before (shaded green), during (shaded red) and after (shaded blue) a simulated heatwave, while panels c and d show the same simulations with green roofs (green line) and cool roofs (blue line). (Source: Tewari et al. (2019) {Open Access})

10.4.2 System Mapping

(Key Skill: Sustainability Science)

Your assigned solution may have a variety of direct and indirect economic, social, and ecological impacts that should also be considered. For example, while the creation or urban forest areas may also improve habitat for native wildlife (ecological impacts), it might also provide some limited recreational opportunities for residents (social impact), or it might create concern about community safety and the potential for increased crime. Similarly, replacing traditional roofs on older buildings may be an easy sell, but it may be harder to convince owners of newer buildings with well-functioning roofs to buy in to the replacement program (economic).

Table 10.1   Urban forest covers surface temperature compared to metro mean of entire area

Region	Daytime temperatures			Nighttime temperatures		
	Metro mean T (°C)	Urban forest T reduction (°C)	<1 km from veg. T reduction (°C)	Metro mean T (°C)	Urban forest T reduction (°C)	<1 km from veg. T reduction (°C)
Cairo	38.8	−5.5	−0.6	26.5	−0.8	−0.5
Delhi	38.4	−1.8	−2.0	25.6	−0.6	0.5
Dubai	46.4	−5.2	0	26.2	−0.7	−1.9
La Paz	32.0	−3.6	−1.3	3.0	1.0	0.8
Lima	26.8	−2.4	−1.9	14.6	0.1	0.1
Los Angeles	37.1	−4.2	−0.9	17.7	−0.7	0.8
Madrid	39.5	−6.0	−0.2	23.4	1.4	1.5
Mexico City	32.3	−3.6	+0.5	16.9	1.4	1.8
Phoenix	44.3	−16.5	−8.7	25.8	−0.1	−0.6
Tehran	39.8	−7.0	−2.2	22.6	0.2	0.1

Source: Dialesandro et al. 2019 (CC *BY* 3.0)

Fig. 10.8   Template sustainability map to help connect ecological, societal, and economic considerations associated with your potential solutions

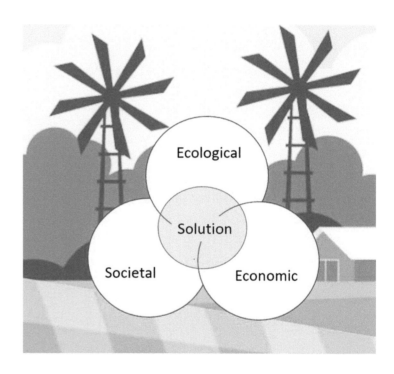

Working with other class members assigned the same solution, develop a simple sustainability map that shows the various system connections across each of the three sustainability domains (ecological, economic, and societal). When you find connections between impacts that cross domains, draw a line to identify the connection (e.g., increased property values because of new roofs (economic) leading to increased financial resources for a family (social)).

The goal is to think broadly about this larger system and envision how implementing actions in one domain (e.g., ecological) may impact components in another domain (e.g., economic or societal).

Use the following template (Fig. 10.8) to get started:

When each group has completed their basic sustainability map, **come together as a class to compare maps for the three possible solutions.** This information will help inform your decision support work later in the unit.

Unit Analysis Summary   Based on your explorations, what have you learned that can help inform your choice of a solution? Do the data support the adoption of one or more than one of these potential solutions?

10.4.3 Reflecting on Your Work

(Key Skill: Personal Reflection)

During your work in Analysis, you explored some of the research into possible solutions to help inform your decision. Take a moment to reflect on this work. Consider the following prompts but feel free to expand on any to best capture your learning experience and better inform your next steps.

- How did you feel working with data? Do you consider quantitative literacy a strength or an area for improvement for you?
- How important should science be in informing management and policy? Do you feel the data you examined support and justify the costs of addressing your challenge?
- Any solution should be examined using a sustainability science lens. Your goal is to find a way to mitigate the UHI effect in this city. But any solution you implement may have other direct and indirect impacts. What are other possible economic, social, or ecological impacts? How should these considerations influence your decision?

10.5 Urban Heat Islands: Solutions

Specific Skills You'll Need to Review: Problem-Solving, Decision Support, Communicating Science

Review your Unit Challenge and the major findings from Discovery and Analysis, including the sustainability map you created to highlight connections among possible solutions and the larger socioecological system.

10.5.1 Small Group Guided Worksheets

(Key Skill: Decision Support)

Decision support matrices can help break down the desired outcomes to reflect multiple criteria for consideration, and they allow you to compare how each possible solution achieves those desired outcomes. This not only helps inform decision making, but it provides transparency in the decision process and justification to help you advocate for adopting a particular solution.

For the three possible solutions, you will evaluate how well each can achieve the following desired outcomes:

- Reduces the city's UHI effect
- Provides long-term reductions in cooling energy needs (proxy for GHG emissions) given a changing climate
- Minimizes costs
- Provides secondary environmental, social, or economic benefits
- Is widely acceptable to stakeholder groups

Considering your three potential solutions, develop a formal decision support matrix to compare and evaluate each approach using the template matrix below (Table 10.2). Getting the most out of the decision matrix requires a depth of knowledge about each of the three possible solutions. Below we list additional sources about each option. Please review each of these, paying particular attention to the two alternatives to your assigned approach.

Note that your group may have uncovered a novel solution not included in this list of three. You may choose to work through the structured decision matrix with your self-identified solution as a fourth solution option.

10.5.2 Additional Sources

1. Shade trees: Shade Trees Reduce Impacts From Urban Heat Islands | by Sabriga Turgon |Medium
2. Green roofs: Using Green Roofs to Reduce Heat Islands | US EPA
3. "Cool roof" reflective surfaces: Increasing Heat Resilience in the Built Environment with Cool Roofs, Cool Walls, and Cool Pavements

Score each of your three potential solutions for each of the desired outcomes using a simple relative scale of 1 for least benefit to 5 for greatest benefit. Using a relative scale means you don't need to know exactly how well each solution meets the goal of each desired outcome, but you can use your judgment to assess how well each solution works compared to the others.

While this is a relative (subjective) scale, note that you will need to justify your scoring of each solution for each desired outcome.

Once each cell in your decision matrix has a relative score, calculate an average score for each solution. Based on this analysis, which is the "best" solution, considering all your desired outcomes?

10.5.3 Role Playing

(Key Skill: Communicating Science)

In tackling environmental issues, you'll often find yourself working with groups of people with different perspectives about implementing a solution. You should use clear, concise communication to summarize your solution and justify its selection. To be effective, you must address concerns likely to be presented by various stakeholder groups and how the risks of taking these actions outweigh the risks of taking no action.

Table 10.2 Basic decision matrix to compare the three possible solutions for your Unit Challenge. Include your justification for each solution's ranking for each desired outcome

		Desired Outcomes			
Solution Options	Minimizes UHI effect	Reduces energy demand	Minimizes costs	Provides secondary benefits	Maximizes stakeholder buy-in
Green roofs					
Cool roofs					
Urban forests					
Justification for Ranking					

Use a scale of 1-5 to score each option under each desired outcome category where 1 = does not achieve the desired outcome and 5 = completely achieves the desired outcome.

Your group will present and justify your chosen solution to the class with a particular emphasis on how it might benefit or impact one of the key stakeholder groups listed below. Be sure to include arguments that support this solution from the ecological, societal, and economic domains of sustainability science. Listeners will also be **assigned a stakeholder identity**, with an opportunity to **ask follow-up questions** after your presentation that reflect their unique concerns and perspectives. Your job will be to listen carefully and tailor your answers to this audience of stakeholders.

Key stakeholder groups include the following:

- Local residents who live and work in the metro area buildings
- Building owners
- The mayor and city planners
- Representatives of a local climate action group
- An energy analyst working for the US DOE

10.5.4 Reflecting on Your Work

(Key Skill: Personal Reflection)

During your work in Solutions, you have explored several possible actions that could be taken to mitigate the impacts of the UHI effect in your city. Take a little more time now to reflect on your findings and the skills you practiced. Consider

the following prompts but feel free to expand on any of them to best capture your learning experience and feelings about this issue.

- Reflect on your work through a sustainability science lens. Does your solution address ecological, economic, and societal considerations? Which considerations do you think should carry the most weight in the decision? Why?
- How did you weigh solutions that might have the greatest environmental impact against those that are most likely to be implemented and maintained for long-term impact?
- How can environmental scientists work to show the value of healthy ecosystems and justify the costs of mitigation strategies?
- Science communication can be challenging, especially when working with diverse audiences. We need to craft our communication to match the interests and values of the target audience, but how do you do this when your audience contains a mix of stakeholder groups? How can you maximize the impact of your message to a diverse audience?

Unit Solution Summary *Summarize and justify your final solution choice and outline how it addresses the direct challenge while also considering social, economic, and ecologi-*

cal impacts. Also demonstrate that it will continue to meet the challenges posed by climate change.

10.6 Urban Heat Islands: Final Challenge

As a part of this Unit Challenge, you were asked to write a one-page Fact Sheet justifying your choice for an approach to mitigate the impacts of the UHI effect in your city. Your Fact Sheet should include the following components:

- Brief problem statement
- Recommended mitigation strategy with sufficient details to summarize the general approach
- Justification of this recommendation (e.g., long-term effectiveness, given anticipated climate changes, implementation costs, other benefits provided, etc.). Be sure to use a sustainability lens to include considerations of direct and indirect ecological, social, and economic considerations
- Any obstacles the group might face trying to implement your solution

Consider the use of figures, graphics, and tables to help summarize the system and how this solution is well suited to meet all desired outcomes.

Final Unit Challenge Submit your final recommendations in a one-page Fact Sheet using clear science communication designed for a lay audience.

References

Borunda A (2021) A Shady divide. Nat Geo 240(1):66–83

Dialesandro JM, Wheeler SM, Abunnasr Y (2019) Urban heat island behaviors in dryland regions. Environ Res Commun 1(8): 081005

International Energy Agency (2018) The future of cooling. IEA, Paris. https://www.iea.org/reports/the-future-of-cooling

Ko Y, Lee J-H, Gregory McPherson E, Roman LA (2015) Long-term monitoring of Sacramento Shade Program trees: tree survival, growth, and energy-saving performance. Landsc Urban Plan 143:183–191

Levinson RM, Gilbert HE, Pomerantz M, Harvey JT, Ban-Weiss GA (2017) Recent cool pavement research highlights: quantifying the energy and environmental consequences of cool pavements. Heat Island Group, Berkeley Lab

Park J, Kim J, Lee DK, Park CY, Jeong SG (2017) The influence of small green space type and structure at the street level on urban heat island mitigation. Urban For Urban Green 21:203–212

Santamouris M (2020) Recent progress on urban overheating and heat island research. Integrated assessment of the energy, environmental, vulnerability and health impact. Synergies with the global climate change. Energy Build 207:109482

Shickman K, Rogers M (2020) Capturing the true value of trees, cool roofs, and other urban heat island mitigation strategies for utilities. Energy Effic 13:407–418

Tewari M, Yang J, Kusaka H, Salamanca F, Watson C, Treinish L (2019) Interaction of urban heat islands and heat waves under current and future climate conditions and their mitigation using green and cool roofs in New York City and Phoenix, Arizona. Environ Res Lett 14:034002

U.S. Environmental Protection Agency (2008) Urban heat Island basics. In: Reducing urban heat islands: compendium of strategies. Draft. https://www.epa.gov/heat-islands/heatisland-compendium

Wahba SM, Kamel BA, Nassar KM, Abdelsalam AS (2018) Effectiveness of green roofs and green walls on energy consumption and indoor comfort in arid climates. Civil Eng J 4(16):2284–2295

Core Knowledge

Sustainable agriculture, Carbon sequestration, Soil science, Food systems

transition. **In this unit, you'll investigate several regenerative agriculture techniques that can reduce the loss of carbon from soils.**

11.1 Environmental Issue

Pressures to feed an ever-growing global population have led to the widespread adoption of "industrial agriculture," the large-scale, intensive production of crops and animals, often relying on soil tilling and frequent applications of chemical fertilizers. In addition to the many environmental impacts of such high-intensity activities, industrial agricultural practices also release greenhouse gases (GHGs) like CO_2 and N_2O to the atmosphere. The global food production system accounts for more than 15% of all human-related GHGs.

Sustainable farming practices that minimize impacts on soil systems and surrounding ecosystems have long been employed at smaller scales. However, regenerative agricultural practices (Fig. 11.1) are being proposed as a means of food production that can also improve the environment. In particular, scientists, farmers, and others in the agricultural community have begun to explore how soils can contribute to climate mitigation by sequestering large quantities of CO_2, the opposite of conventional agriculture. With agricultural uses covering more than a third of the Earth's land surface, it is estimated that by employing regenerative agricultural practices, soil carbon sequestration could remove up to 20 gigatons of CO_2 over the next several decades.

While several high-profile agricultural companies are pledging to advance regenerative agriculture, and the Regenerative Organic Alliance has established a certification program, widespread adoption is still a ways off. Only 1% of agriculture in the United States is organic, and an even smaller percentage is regenerative. To bring regenerative agricultural practices into greater use, region-specific practices must be investigated and adopted, with technical and financial support to help farming communities make the

11.2 Background Information

11.2.1 The Problem

Modern agriculture is beset by a host of environmental problems, ranging from human health threats from exposure to pesticides and herbicides to pollution caused by the runoff of sediments into nearby waterways. Impacts can be localized in nature, as is the case with exposure of agricultural workers to pesticide spray, or span thousands of miles, as is the case with the annual formation of the Gulf of Mexico Dead Zone fueled by runoff of nitrate fertilizers from the Mississippi River Basin. These practices also impact the soil system itself, with more than 20% of the world's soils experiencing various degrees of degradation (Fig. 11.2). However, agriculture also has significant impacts on climate change. The USDA estimates that agriculture accounted for 11.2 % of US GHG emissions in 2020, primarily from cropped and grazed soils.

Globally, soils store three times more carbon than is present in the atmosphere and four times the amount in living plants and animals, making it a significant carbon sink. However, current agricultural and land management practices, including tilling and removing crop residues and excessive applications of pesticides and fertilizers, are turning soils into a significant carbon source rather than a sink. In addition, intensive use of synthetic nitrogen fertilizers has led to significant releases of N_2O, a powerful GHG.

Recent estimates indicate that emissions from soils have accounted for about a quarter of all human-related GHG emissions. Because nearly half of the globe's arable land now supports pastures, rangelands, and croplands, soils have lost 50–70% of the carbon they once stored (Cho 2018).

J. Pontius, A. McIntosh, *Environmental Problem Solving in an Age of Climate Change*, Springer Textbooks in Earth Sciences, Geography and Environment, https://doi.org/10.1007/978-3-031-48762-0_11

Fig. 11.1 Regenerative agriculture in practice. (Source: Luigi Guarino, CC BY 2.0 via Wikimedia Commons)

Fig. 11.2 Overview of the state of global soil degradation in the world. (Source: Philippe Rekacewicz, UNEP/ GRID-Arendal www.grida.no/ resources/5507)

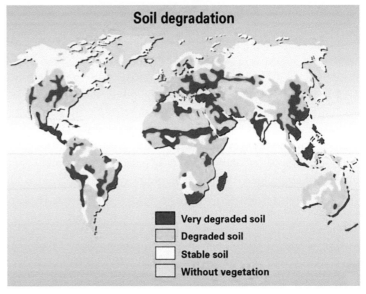

However, such losses are not inevitable. Proper soil management can regenerate a soil's ability to sequester and store carbon.

Scientists believe that soils could continue to store carbon for another 20–40 years before they become saturated. Given the added benefit that stored carbon can also lead to improved soil health and agricultural productivity, better soil management, including carbon storage, seems like a wise move. However, the challenge of having agricultural land becomes a carbon sink rather than a carbon source while still meeting the needs of an ever-growing human population remains.

11.2.2 The Role of Climate Change

Agricultural practices can create soils that act as either a significant source of GHG emissions or a significant carbon sink. Soils naturally store carbon when CO_2 is removed from the atmosphere by plants, which leave a stable form of soil organic carbon (SOC) when they decompose. However, agricultural practices that disturb the soil structure and alter soil chemistry and microbial communities can cause a net release of the bound SOC. Considering the size of the SOC pool and the proportion of land currently managed for agricultural uses, whether soils act as a carbon source or a sink will have a significant impact on climate change.

Agriculture impacts climate change in many other ways. Fossil fuel inputs are needed by industrial agriculture to plant, fertilize, and process raw materials into consumer-ready products and distribute the finished products around the globe. Cattle and other livestock emit the potent GHG methane (CH_4) from their digestive tracts, and food waste releases CH_4 as it decomposes.

However, climate impacts agriculture as well. Atypical weather like a lack of snow cover in winter can lead to reduced yields of winter cash crops in cold climates, while abnormally hot summers and very cold winters can reduce yields of sensitive crops. Extended drought or flooding rainfalls not only threaten crop survival but can erode and further degrade soils. Warmer temperatures also increase microbial activity, which speeds decomposition and the release of SOC pools.

It's important to note the cyclical nature of some of these impacts. For instance, the release of more GHGs from warming soils will likely speed the warming process, resulting in even greater releases of GHGs from soils as the feedback loop continues to operate.

11.2.3 Solutions

Agricultural lands are unique in that they can be managed in ways that make them either sources of or sinks for GHGs. Some traditional industrial farming techniques (tilling, over fertilizing, over grazing) are designed to maximize short-term productivity without concern for the long-term fertility of the soils or impacts on the larger system. However, regenerative agriculture focuses on the interconnections of farming systems and the ecological system as a whole, with the goal of rebuilding soil organic matter (SOM) (thus sequestering carbon) and restoring degraded soil biodiversity (improving soil and ecosystem function) (Newton et al. 2020). The beauty of regenerative agriculture is that it actually improves the land and restores its natural ability to produce high quality food while simultaneously improving soil quality, sequestering carbon (Fig. 11.3), and ultimately leading to healthy communities and economies.

While exact techniques vary, regenerative agriculture's underlying principles include the following:

1. Maintaining continuous vegetation cover on the soil as much as possible
2. Reducing soil disturbance to promote stabilization of organic matter in soil mineral complexes
3. Increasing the amount and diversity of organic residues returned to the soil
4. Maximizing nutrient and water use efficiency by plants

Fig. 11.3 Latest USDA Natural Resources Conservation Service (NRCS) science and technology help agriculture mitigate climate change. (Source: USDA NRCS {Public Domain})

5. Restoring microbial life essential to soil health and biodiversity

Broadly speaking, these principles are designed to more closely mimic a comparable native ecosystem. Farmers can accomplish this using several techniques, including the following:

- Leaving crop residues after harvest or planting cover crops to protect soil from erosion.
- Rotating crops to include nitrogen-fixing legumes to add nutrients to the soil.
- Using regenerative grazing techniques for livestock to control the amount of time they graze in one area, allowing grasses to regenerate, and promoting soil carbon levels, better water retention, and increased biodiversity.
- Increasing the ability of soils to retain water to dampen the impacts of both floods and droughts. Simply by increasing levels of SOM, farmers can dramatically improve the ability of those soils to hold water.

In this unit, you'll focus on three additional ways that regenerative agriculture improves soil health, sequesters and stores carbon, and helps slow the progress of climate change:

- Adding carbon-containing amendments to soils
- Minimizing tillage as a way to prepare soils for planting (Fig. 11.4)
- Replacing synthetic nitrogen fertilizers with green manure or organic fertilizers

11.2.4 Unit Challenge

As the Chief Soil Scientist working in your state's USDA NRCS Soil and Plant Science Division, you've been asked to evaluate the best way to promote soil management techniques to help meet GHG emission reduction goals put forth

Fig. 11.4 David Schroeder farms about 1200 acres of soybeans and corn in Iowa using no-till techniques. (US EPA {Public Domain})

Fig. 11.5 Biochar is an evolved charcoal made from woody biomass. Current uses include soil amendment, stormwater filtration, and environmental remediation. (Source: OR Department of Forests(CC BY 2.0) via Flickr)

Fig. 11.6 No-till agricultural practices minimize soil disturbance and provide continuous cover with crop residue. (Source: USDA NRCS South Dakota {Public Domain})

in the state's strategic plan. Specifically, your Unit Challenge is to evaluate how regenerative agriculture could be adopted by local farmers to best accomplish the following: (1) reduce GHG levels in the atmosphere, (2) improve the health and biodiversity of soils, and (3) maintain high crop yields.

11.2.5 The Scenario

After reviewing the various regenerative agriculture techniques that might be appropriate for your particular state and available data on local agroecosystems, you decide to focus on the following three approaches:

1. **Biochar amendments.** Capturing carbon contained in organic material and adding it to soil can both improve soil quality and lower GHG levels in the atmosphere. Carbon-capturing soil amendments include sugarcane bagasse and biochar, a charcoal-like substance that's made by burning organic material from agricultural and forestry wastes in a controlled process called pyrolysis (Fig. 11.5).

During pyrolysis, organic materials, such as wood chips, leaf litter, or dead plants, are burned in a container with very little oxygen. During the process, organic material is converted into biochar, a stable form of carbon that can't easily escape into the atmosphere. The energy or heat created during pyrolysis can be captured and used as a form of clean energy. Biochar is highly efficient at converting carbon into a stable form and is cleaner than other forms of charcoal.

   Specific recommendation: Work with farmers to add biochar to 7500 ha of actively farmed land.

2. **No-till farming.** One of the ways soils are commonly prepared for planting is tilling, which includes digging, stirring, and overturning soils. Tilling stirs up soil humus, speeding the loss of soil carbon to the atmosphere. However, many crops can be successfully planted without tilling the soil (Fig. 11.6). One technique employs the use of implements like the no-till drill, a tool that plants seeds without disturbing the overlying soil. To manage weeds around the planted crops, a variety of methods are used, including free-range livestock and tractor implements such as the roller crimper, which farmers can use to lay down a weed-suppressing mat that can be planted through in one pass.

Fig. 11.7 Spreading manure. (Source: Photo by Brian Forbes {CC BY 2.0} via Wikimedia Commons)

Specific recommendation: Work with farmers to move to no-till practices on 7500 ha of actively farmed land.

3. **Organic fertilizers**. The use of the synthetic fertilizer ammonium nitrate contributes to climate change in several ways. Its manufacture is energy intensive, resulting in substantial releases of GHGs during production. After farmers apply N fertilizers to their crops, N_2O, a potent GHG, is produced and released to the atmosphere. There are several alternatives available to farmers. Planting cover crops such as legumes can increase soil N levels through biological fixation. Organic fertilizers, including compost, plant residues, or manure (Fig. 11.7), can be plowed into soils to provide N and reduce the need to apply expensive synthetic N fertilizers.

Specific recommendation: Work with farmers to replace chemical fertilizers with compost and manure additions on 7500 ha of actively farmed land.

11.2.6 Relevant Facts and Assumptions

- Your goal is to apply each of these solutions on 7500 ha of active farmland.
- These test farms have been previously used for corn production using conventional practices.
- The recommended maximum biochar rate of 10 tons per ha costs about $12 per ha.
- The average cost of renting a no-till drill is $31 per ha.
- Composted plant waste is recommended for application at 25 tons per ha; a ton costs $18.
- The predominant soil type in the region is well-drained loam with an average pH of 6.8.

Each of these three approaches has important advantages and disadvantages and costs and benefits. Your task is to evaluate each and recommend the one that will, in your opinion, provide the maximum GHG reduction over time, while minimizing costs and maximizing soil health, crop yield, and community buy in.

11.2.7 Build Your Foundational Knowledge

Below are web sources that provide additional information about each of the solutions you're considering for this Unit Challenge. This information provides a critical foundation to help you evaluate each option and support your final choice. After reviewing each source, be prepared to answer questions in the Preparation Assessment Quiz and to summarize any information relevant to your Unit Challenge.

Biochar amendments:
Soil amendments and inoculants, Biochar basics

No-till farming:
What is no-till farming? Advantages and disadvantages of no-till farming

Organic fertilizers:
Pros and cons of organic vs. chemical fertilizers, Plant residue

Final Product: A one-page Fact Sheet summarizing your proposed regenerative farming technique to promote among farmers in order to help meet the state's GHG targets, as well as justifying its implementation as a part of the larger USDA goals for the state's agricultural sector. Consider your audience, local farmers, town officials and local community members, and be sure to demonstrate how your proposed solution will stand up to the challenges posed by climate change.

11.2.8 Preparation Assessment Quiz

Are you ready to tackle your challenge? At this point you should understand the basic environmental principles and ecological processes involved in this environmental problem. Consider the following questions. If you are comfortable with answering these, then you are ready to head into Discovery, Analysis, and Solutions activities.

- List several ways that intensive or industrial agriculture degrades the environment.
- What are some of the important sources of GHGs released by industrial agriculture?
- What is regenerative agriculture?
- What specific components of regenerative agriculture focus primarily on GHGs?
- What are some of the ways that improved soil management can slow the rate of climate change?
- According to Biochar Basics, what is "Terra Preta"?
- According to the article on the advantages and disadvantages of no-till farming, what are two advantages of no-till?
- According to Green Wing Services Inc., what are three advantages of organic fertilizers?
- For each of the proposed solutions in your Unit Challenge, are there any additional benefits that might arise from their implementation that might not be directly related to the challenge?
- For each of your proposed solutions, are there any negative unintended consequences that might result from their implementation?
- What additional information did you glean from your web sources that might help inform your Unit Challenge?

11.3 Regenerative Agriculture: Discovery

Specific Skills You'll Need to Review: Navigating the Scientific Literature, Science Communication, Problem-Solving

11.3.1 Independent Research

(Key Skill: Navigating the Scientific Literature)

To better understand the potential for each of your regenerative agriculture approaches, you first need to examine the literature to see what others have found. Conduct a search of the peer-reviewed scientific literature focused on the solution you have been assigned, and identify one research paper that focuses on your assigned approach.

Prepare a summary of the article you selected that includes the following:

- **Citation**
- **Main topic**: Stick to a few words, likely pulled from the title.
- **General summary**: A few bulleted sentences summarizing the research question it addresses and approaches it takes.
- **Methods:** How did they approach their research question?

- **Location**: Where was the work done?
- **Conclusions**: Concise list of the findings, specifically capturing the take-home message.
- **Relevance**: How might this study help inform your Unit Challenge? Feel free to make a bulleted list of information you may want to include later.

11.3.2 Literature Share: Reciprocal Instruction

(Key Skill: Scientific Communication)

Share In small groups, share and critique the research article you found. Keep in mind that your peers have not read this article, and it is your job to convey the key information to them. Note the items that will be important to consider when you are developing your solution to the challenge.

Critique Evaluate how these studies might help inform your Unit Challenge. Consider the following:

- Source (Quality of the work or bias of the authors)
- Methods (Did their methods sufficiently address the research question?)
- Conclusions (Did the results justify the conclusions made?)
- Relevance (Can these findings be applied to your challenge?)

Based on your critique, choose one article to share with the larger class, along with the key information that may be useful in deciding on a solution to propose.

11.3.3 Think-Compare-Share

(Key Skill: Problem-Solving)

Now that you have more information about possible solutions for this unit's challenge, you need to **develop a more formal problem definition** to guide your work throughout the rest of the exercises.

Think Start by working independently to develop a specific Problem Statement to guide the remainder of your work. Problem Statements provide the relevant information and boundaries to make the issue something you can effectively assess and tackle. The basics of a formal Problem Statement include the following:

Problem Statement: A short, concise statement summarizing the issue that includes the following:

- A **description** of the undesired condition or change that you hope to achieve (What is the actual problem?)
- **Justification** for addressing the problem (Why does this problem matter?)
- Potential **sources** or **causes** of this problem (What is the cause you need to address?)
- The **metrics** you will use to assess the status of the problem (How will you know if you are making a difference in the problem?)
- The **desired outcome** for these metrics (What is the end goal or ideal state?)
- Potential **solutions** to consider (How might you attempt to achieve this goal?)

Compare/Share Now return to your small group to share your Problem Statements. Use each of your ideas to develop a joint Problem Statement that contains all key information and is concise, clear, and well written.

Unit Discovery Summary *Submit a final Problem Statement that succinctly captures the key information to guide your work on this Unit Challenge.*

11.3.4 Reflecting on Your Work

(Key Skill: Personal Reflection)

After your work in Discovery, you should have a better idea of the problems you face and have produced a Problem Statement you can use to tackle the Unit Challenge. Take a moment to reflect on this work. Consider the following prompts but feel free to expand on any to best capture your learning experience and better inform your next steps.

- Of the skills you practiced in Discovery, which were the most challenging? Which were the most interesting?
- How were you most comfortable working during these exercises? In small groups, independently, or with the larger class? Why? How does your choice reflect your personality type and leadership style?
- Was your Problem Statement strictly focused on the environmental problems linked to industrial agriculture, or did it also consider important social and economic considerations? How might a focus on the environmental aspects limit your ability to identify truly sustainable solutions?
- You've been given three viable solutions to assess as a part of this case study. But this is not an exhaustive list of options or even necessarily the best possible course of action for every scenario. Take a moment to "think outside the box." Are there any other possible solutions you think would be worth exploring? Describe one that you think would be worth pursuing.

11.4 Regenerative Agriculture: Analysis

Specific Skills You'll Need to Review: Quantitative Literacy, Sustainability Science

Review your Background and Discovery sections before beginning the Rotating Station exercises below. While you focused on one potential solution in your Independent Research in Discovery, keep an open mind as your work through Analysis activities.

11.4.1 Rotating Stations

(Key Skill: Quantitative Literacy)

At each of the following stations, you will review data that are relevant to the three potential solutions you're considering. Spend some time working through the analyses at each station to learn more about this issue and possible solutions for your Unit Challenge.

Be sure to write down one finding at each station that will help inform your selection of a solution.

Station 1: Biochar Amendments The addition of biochar to croplands can result in a considerable amount of carbon retention in soils. Shin et al. (2017) analyzed the effectiveness of biochar in sequestering carbon when combined with different types of compost in corn cultivation. They looked at both the amount of carbon sequestered by different combinations of compost and compost plus biochar over the growing season (Fig. 11.8) and the carbon held in the soil and removed from the atmosphere over the full cultivation period (Table 11.1).

After examining the data, answer the following questions:

- Which of the three compost treatments had the highest total soil carbon content over the full cultivation period?
- Which had the greatest increase in carbon sequestration with the addition of biochar?
- How did total carbon content change over time (days since sowing)?
- What is the benefit of adding biochar with compost rather than compost alone?
- Assume that a farmer wants to use compost from his aerobic digester (AD), mixed with biochar, to receive income from the sale of carbon credits. If the farmer can apply this treatment to 650 ha and a carbon credit pays $7.90 per

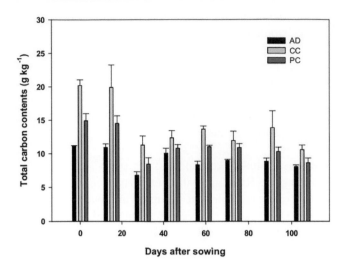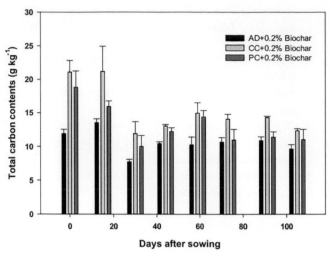

Fig. 11.8 Changes in total carbon content of soil amended with biochar mixed with different compost sources (AD aerobic digestate, CC cow compost, PC pig compost). (Source: Shin et al. 2017)

Table 11.1 Carbon sequestration in the soil supplied with various amendments over the full cultivation period

Treatments[a]	Compost only (A)	With biochar (B)	C-sequestration (B–A)	Recovery rates (D)[b]
	kg ha^{-1}			(%)
AD	10,491	12,532	2041	70.0
CC	13,754	16,055	2301	78.5
PC	11,245	13,221	1976	67.4

Source: Shin et al. (2017)

Biochar input; 2600 kg ha^{-1} (TC; 56.4%)

[a]*AD* aerobic digestate, *CC* cow compost, *PC* pig compost

[b]D = C/Accumulated application amount of area × (TC/100)

MT CO_2, how much additional income can the farmer expect to make from the sale of carbon credits?

Record any relevant Station 1 findings

Station 2: No-Till Agriculture Important sources of GHGs in agriculture are the fuels used to operate the machinery needed to prepare the soils for planting crops. Demir and Gözübüyük (2020) evaluated equipment CO_2 emissions from different types of tillage systems in Turkey. Table 11.2 compares emissions from conventional tillage, two types of reduced tillage, and no-till seed planting for three different crops.

Based on the data in the table above, answer the following about emissions from equipment used for conventional, reduced tillage, and no-till seeding methods for planting:

- Which crop produced the highest total CO_2 emissions?
- What percent decrease in CO_2 emissions could be expected when switching from conventional (CT) to no-till (NT) seeding for each of the three crops?

- In addition to GHG emissions related to fuel and oil use during soil treatment, how else might GHGs be released during the conventional tilling process?
- Assume that a farmer wants to earn income from carbon credits for a switch from conventional tillage to no-till practices with his sunflower crop. If the farmer can make this switch on 650 ha, how many kg of CO_2 emissions from farming equipment can be saved with this change in tilling practices? Assume that there are an additional 1500 kg CO_2 ha^{-1} that are retained in undisturbed soils as a result of this new tilling practice and a carbon credit pays $7.90 per MT CO_2. How much additional income can the farmer expect to make from the sale of carbon credits?

Record any relevant Station 2 findings

Station 3: Organic Fertilizers Composted plant waste added to soils as an organic fertilizer can alter the behavior of N in soils and potentially reduce the release of nitrous oxide (N_2O), a potent GHG, to the atmosphere. Carter et al. (2014) compared several techniques for incorporating plant

Table 11.2 CO_2 emissions (kg CO_2 ha^{-1}) from tillage systems

Treatments	Fuel-based CO_2 emissions			Oil-based CO_2 emissions			Total emissions		
	Vetch	Wheat	Sunflower	Vetch	Wheat	Sunflower	Vetch	Wheat	Sunflower
CT	151.04	142.91	154.69	6.93	6.56	7.10	157.97	149.46	161.79
RT1	76.16	77.13	89.05	3.49	3.54	4.09	79.65	80.66	93.13
RT2	78.05	77.97	93.14	3.58	3.58	4.27	81.63	81.55	97.42
NT	31.63	32.64	30.86	1.45	1.50	1.42	33.08	34.14	32.27

Source: Demir and Gözübüyük (2020)
CT conventional tillage, *RT1* reduced tillage 1, *RT2* reduced tillage 2, *NT* no-till seeding

Table 11.3 Cumulative fluxes of N_2O and CO_2 during a 92-day period measured on soil units amended with different fertilizers (means, SE in brackets; $n = 4$)

Treatment[a]	Cum. N_2O flux (mg N/m^2)	Cum. CO_2 emission (g C/m^2)	% of added C
CON	−1.4 (2.6) ab	62.3 (0.6) b	
MIN	13.2 (3.2) ac	51.6 (0.9) a	
COM-PLO	0.0 (1.9) ab	121.9 (1.0) c	32
SIL-PLO	35.6 (7.5) c	146.6 (1.0) d	47
COM-HAR	−11.8 (2.0) b	146.2 (1.3) d	44
SIL-HAR	16.6 (4.3) ac	160.3 (1.3) e	54

Source: Adapted from Carter et al. (2014)
[a]Flux rates with the same letter are not significantly different

waste/organic fertilizer into soils and their impact on emissions of both CO_2 and N_2O.

Plant wastes, either grass-clover compost (COM) or silage (SIL), were either plowed into the soil to a depth of 15 cm (PLO) or harrowed to a depth of 5 cm (HAR). Table 11.3 compares N_2O and CO_2 releases from the various treatments.

Use the information in the table to answer the following questions:

- Which treatment had the lowest N_2O flux from the soil? Which treatment had the lowest CO_2 emissions?
- Note that the last column in the table shows cumulative CO_2 emissions from the soil as a percent of carbon added (sequestered) in the soil. This sequestered carbon can be considered an offset to CO_2 emissions since it is bound and held by organic matter in the soil. Start by converting the percent of added carbon to g C m^{-2} for each treatment. With this information you can now determine whether or not your soils are a net carbon source (higher emissions than sequestration) or net carbon sink (higher sequestration than emissions). Which treatment has the lowest net carbon flux when offsetting emissions with these carbon additions to soils?
- How does this net carbon flux (considering carbon sequestered in soils) compare to the control and mineral fertilizer treatments (assume 0% added C for the CON and MIN treatments)?

Record any relevant Station 3 findings

Station 4 Tiefenbacher et al. (2021) published a synthesis of research exploring carbon sequestration potential for a range of agricultural practices, including no-till farming, organic fertilizers (manure, compost and crop residue), and a control (bare fallow).

Study Fig. 11.9 and answer the following questions:

- Considering the 95% confidence intervals in this bar and whiskers chart, which treatments can be expected to sequester carbon (significantly greater than 0 at the 0.05 alpha-level)?
- Which treatment has the greatest carbon sequestration potential? Is this significantly better than any of the other treatments at the 0.05 alpha-level?
- For your challenge, you are comparing biochar additions, no-till practices, and organic fertilizer additions. Assume the biochar has a carbon sequestration potential of 250 kg C/ha/year. Use the median values in Fig. 11.9 to estimate the carbon sequestration potential for no-tillage and compost treatments. Which treatment has the potential to sequester the most carbon?

Armed with the answers to the questions above, farmers might choose to use a secondary factor (e.g., cost of application) to determine which method to use. For your Unit Challenge, you were given a set of Relevant Facts and Assumptions to help guide your decisions. Use that information to complete the following:

- Calculate the cost to apply each of these techniques to 7500 ha of active farmland. Which would be the cheapest?
- Now calculate the cost per kg C -ha -yr for each of these treatments. Which treatment would provide the lowest cost per kg of C sequestered in a year?
- If money were not a factor, which treatment would you choose? If you were limited to a $100,000 investment, which treatment would you choose?

Record any relevant Station 4 findings

Fig. 11.9 Carbon sequestration potential across various regenerative agriculture techniques with 95% confidence intervals. (Source: Tiefenbacher et al. 2021 {CC-BY-4.0})

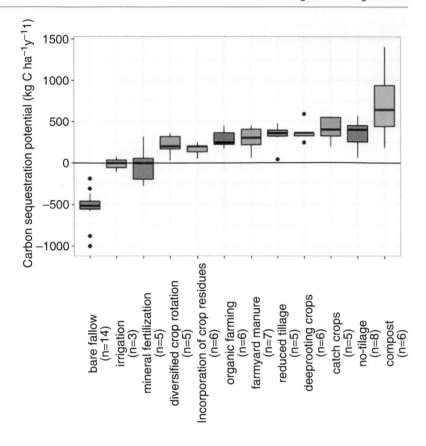

11.4.2 System Mapping

(Key Skill: Sustainability Science)

Your assigned solution may have a variety of direct and indirect economic, social, and ecological impacts that should also be considered. For example, rental of no-till equipment may increase profits for local farm supply businesses, while adding organic fertilizers may improve soil health long term (e.g., increased water-holding capacity).

Working with other class members assigned the same solution, develop a simple sustainability map that shows the various system connections across each of the three sustainability domains (ecological, economic, and societal). When you find connections between impacts that cross domains, draw a line to identify the connection (e.g., improved soil biodiversity (ecological) leading to greater soil fertility and increased crop yields (economic)).

The goal is to think broadly about this larger system and envision how implementing actions in one domain (e.g., ecological) may impact components in another domain (e.g., economic or societal).

Use the following template (Fig. 11.10) to get started:

When each group has completed their basic sustainability map, **come together as a class to compare maps for the three possible solutions.** This information will help inform your decision support work later in the unit.

Unit Analysis Summary Based on your explorations, what have you learned that can help inform your choice of a solution? Do the data support the adoption of one or more than one of these potential solutions?

11.4.3 Reflecting on Your Work

(Key Skill: Personal Reflection)

During your work in Analysis, you explored some of the research into possible solutions to help inform your decision. Take a moment to reflect on this work. Consider the following prompts but feel free to expand on any to best capture your learning experience and better inform your next steps.

- How did you feel working with data? Do you consider quantitative literacy a strength or an area for improvement for you?
- How important should science be in informing management and policy? Do you feel the data you examined support and justify the costs of addressing your challenge?
- Any solution should be examined using a sustainability science lens. Your goal is to use regenerative agriculture techniques to sequester CO_2 and improve soil health. But any solution you implement may have other direct and

Fig. 11.10 Template sustainability map to help connect ecological, societal, and economic considerations associated with your potential solutions

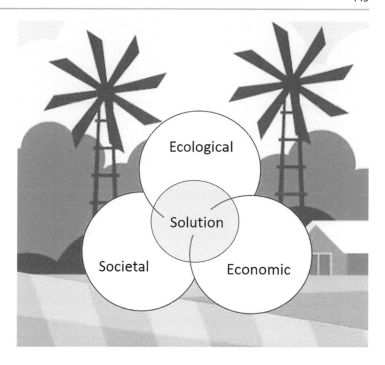

indirect impacts. What are other possible economic, social, or ecological impacts? How should these considerations influence your decision?

11.5 Regenerative Agriculture: Solutions

Specific Skills You'll Need to Review: Problem-Solving, Decision Support, Communicating Science

Review your Unit Challenge and the major findings from Discovery and Analysis, including the sustainability map you created to highlight connections among possible solutions and the larger socioecological system.

11.5.1 Small Group Guided Worksheets

(Key Skill: Decision Support)

Decision support matrices can help break down the desired outcomes to reflect multiple criteria for consideration, and they allow you to compare how each possible solution achieves those desired outcomes. This not only helps inform decision-making, it provides transparency in the decision process and justification to help you advocate for adopting a particular solution.

For the three possible solutions, you will evaluate how well each can achieve the following desired outcomes:

- Effectively sequesters C in the soil
- Improves soil health with a long-term impact given a changing climate
- Minimizes costs
- Provides secondary environmental, social, or economic benefits
- Is widely acceptable to stakeholder groups

Considering your three potential solutions, develop a formal decision support matrix to compare and evaluate each approach using the template matrix below (Table 11.4). Getting the most out of the decision matrix requires a depth of knowledge about each of the three possible solutions. Below we list additional sources about each option. Please review each of these, paying particular attention to the two alternatives to your assigned approach.

Note that your group may have uncovered a novel solution not included in this list of three. You may choose to work through the structured decision matrix with your self-identified solution as a fourth solution option.

11.5.2 Additional Sources

1. Biochar amendments: Biochar for climate change: Is it a viable strategy? | Farming Connect (gov.wales)
2. No-till farming: Could No-Till Farming Reverse Climate Change? (usnews.com)
3. Organic fertilizers: Organic Agriculture Helps Solve Climate Change | NRDC

Table 11.4 Basic decision matrix to compare the three possible solutions for your Unit Challenge. Include your justification for each solution's ranking for each desired outcome

	Desired Outcomes				
Solution Options	Maximizes C sequestration	Long-term soil health	Minimizes costs	Provides secondary benefits	Maximizes stakeholder buy-in
Biochar amendments					
No-till practices					
Organic fertilizer					
Justification for Ranking					

Use a scale of 1-5 to score each option under each desired outcome category where 1 = does not achieve the desired outcome and 5 = completely achieves the desired outcome.

Score each of your three potential solutions for each of the desired outcomes using a simple relative scale of 1 for least benefit to 5 for greatest benefit. Using a relative scale means you don't need to know exactly how well each solution meets the goal of each desired outcome, but you can use your judgment to assess how well each solution works compared to the others.

While this is a relative (subjective) scale, note that you will need to justify your scoring of each solution for each desired outcome.

Once each cell in your decision matrix has a relative score, calculate an average score for each solution. Based on this analysis, which is the "best" solution, considering all your desired outcomes?

11.5.3 Role Playing

(Key Skill: Communicating Science)

In tackling environmental issues, you'll often find yourself working with groups of people with different perspectives about implementing a solution. You should use clear, concise communication to summarize your solution and justify its selection. To be effective, you must address concerns likely to be presented by various stakeholder groups and how the risks of taking these actions outweigh the risks of taking no action.

Your group will present and justify your chosen solution to the class with a particular emphasis on how it might ben-

efit or impact one of the key stakeholder groups listed below. Be sure to include arguments that support this solution from the ecological, societal, and economic domains of sustainability science. Listeners will also be **assigned a stakeholder identity**, with an opportunity to **ask follow-up questions** after your presentation that reflect their unique concerns and perspectives. Your job will be to listen carefully and tailor your answers to this audience of stakeholders.

Key stakeholder groups include the following:

- Local farmers who will implement regenerative agricultural practices
- Agricultural Extension agents who will provide assistance to farmers
- State environmental officials working on agriculture-related issues
- NGOs focused on reducing the impacts of climate change on the state
- Consumer groups concerned about crop prices

11.5.4 Reflecting on Your Work

(Key Skill: Personal Reflection)

During your work in Solutions, you have explored several regenerative agriculture steps that could be taken to reduce GHG emissions in your state. Take a little more time now to reflect on your findings and the skills you practiced. Consider the following prompts but feel free to expand on any of them to best capture your learning experience and feelings about this issue.

- Reflect on your work through a sustainability science lens. Does your solution address ecological, economic, and societal considerations? Which considerations do you think should carry the most weight in the decision? Why?
- How did you weigh solutions that might have the greatest environmental impact against those that are most likely to be implemented and maintained for long-term impact?
- How can environmental scientists work to show the value of healthy ecosystems and justify the costs of mitigation strategies?
- Science communication can be challenging, especially when working with diverse audiences. We need to craft our communication to match the interests and values of the target audience, but how do you do this when your audience contains a mix of stakeholder groups? How can you maximize the impact of your message to a diverse audience?

Unit Solution Summary *Summarize and justify your final solution choice and outline how it addresses the direct challenge while also considering social, economic, and ecological impacts. Also demonstrate that it will continue to meet the challenges posed by climate change.*

11.6 Regenerative Agriculture: Final Challenge

As a part of this Unit Challenge, you were asked to prepare a one-page Fact Sheet summarizing your proposed regenerative agriculture technique to promote among farmers in order to help meet the state's GHG targets, as well as justifying its implementation to the state's agricultural community. Your Fact Sheet should include the following components:

- Brief problem statement
- Recommended mitigation strategy with sufficient details to summarize the general approach
- Justification of this recommendation (e.g., long-term effectiveness, given anticipated climate changes, implementation costs, other benefits provided, etc.). Be sure to use a sustainability lens to include considerations of direct and indirect ecological, social, and economic considerations
- Any obstacles the group might face trying to implement your solution

Consider the use of figures, graphics, and tables to help summarize the system and how this solution is well suited to meet all desired outcomes.

Final Unit Challenge Submit your final recommendations in a one-page Fact Sheet using clear science communication designed for a lay audience.

References

Carter MS, Sørensen P, Petersen SO, Ma X, Ambus P (2014) Effects of green manure storage and incorporation methods on nitrogen release and N_2O emissions after soil application. Biol Fertil Soils 50:1233–1246. https://doi.org/10.1007/s00374-014-0936-5

Cho R (2018) Can soil help combat climate change? Columbia Climate School State of the Planet. https://news.climate.columbia.edu/2018/02/21/can-soil-help-combat-climatechange/

Demir O, Gözübüyük Z (2020) A comparison of different tillage systems in irrigated conditions by risk and gross margin analysis in Erzurum region of Turkey. Environ Dev Sustain 22:2529–2544. https://doi.org/10.1007/s10668-019-00308-5

Newton P, Civita N, Frankel-Goldwater L, Bartel K, Johns C (2020) What is regenerative agriculture? A review of scholar and practitioner definitions based on processes and outcomes. Front Sustain Food Syst 4: 577723

Shin J, Hong SG, Lee S, Hong S, Lee J (2017) Estimation of soil carbon sequestration and profit analysis on mitigation of CO_2-eq. emission in cropland cooperated with compost and biochar. Appl Biol Chem 60(4):467–472

Tiefenbacher A, Sandén T, Haslmayr HP, Miloczki J, Wenzel W, Spiegel H (2021) Optimizing carbon sequestration in croplands: a synthesis. Agronomy 11(5):882

Core Knowledge

Tropical ecology, Forest ecology, Ecosystem services, Natural resource management, Deforestation

12.1 Environmental Issue

The Amazon rainforest covers more than 2 million square miles or about 518 million hectares (ha) of northern South America and has long been recognized as a critically important biodiversity hotspot. This expansive biome is home to an estimated 30% of the world's plant and animal species, including 2500 species of trees alone. In addition to providing critical habitat, the rainforest is home to more than 30 million people, including about 350 indigenous groups which depend on the forest for the resources it provides. However, the rainforest also benefits populations across the globe; almost a quarter of all pharmaceuticals contain ingredients derived from the rainforest.

The Amazon rainforest also plays an important role in weather and climate. Aside from producing almost half of its own rainfall due to evapotranspiration, it also generates precipitation for critical dry regions across South America, supporting agricultural production. The Amazon also has a significant role in the global carbon cycle, with the potential to act as either a carbon source or sink, with implications for global climate trends.

While deforestation in the Amazon has been an ongoing concern for decades, many experts now believe we are approaching a tipping point. Unchecked agricultural expansion, mining, and logging are pushing this biome to the point where it may no longer be able to sustain itself (e.g., continue to produce the rainfall needed to maintain its current structure) (Fig. 12.1). Many predict a "state shift" that will result in a vast area of tropical grassland and the corresponding loss of the many ecosystem services the forest provides. **In this unit, you'll consider ways to protect and preserve the Amazon rainforest.**

12.2 Background Information

12.2.1 The Problem

Globally, tropical forests are disappearing at a rate of about 13 million ha per year, an area approximately the size of Greece. Deforestation and degradation have reduced the area covered by tropical forests from 12% to 5% of the Earth's land area (Brandon 2014). The primary cause is commercial agriculture, but recent data indicate that the loss of tropical forests is increasingly the result of climate change-driven droughts, fires, and pest outbreaks. The resulting loss of ecosystem services these forests provide has impacts both locally and globally.

The world's largest rainforest, the Amazon rainforest of Brazil, is in particular trouble. Widely recognized as a biodiversity hotspot and key regulator of both regional weather patterns and global climate, the forest has been transformed by a host of human activities. Logging for timber and wood products. clearing for agriculture (Fig. 12.2), mining operations, hydroelectric projects, and infrastructure (e.g., roads) to access remote areas are all major threats to the forests that remain.

Deforestation and climate change may already have pushed the Amazon close to a critical threshold of dieback from which it cannot recover. Examining satellite data from over the past two decades, Boulton et al. (2022) reported that nearly 75% of the rainforest has lost resilience, or the ability to recover and restore baseline structure and function after disturbance. This inability to recover from disturbances like fires and logging and the ongoing stress caused by climate change may push the rainforest past a tipping point, resulting in the forest becoming a grassy savannah with reduced capacity to generate rainfall, store carbon, or produce the suite of other ecosystem services on which we depend. However, when we may reach this tipping point remains uncertain.

J. Pontius, A. McIntosh, *Environmental Problem Solving in an Age of Climate Change*, Springer Textbooks in Earth Sciences, Geography and Environment, https://doi.org/10.1007/978-3-031-48762-0_12

Fig. 12.1 2020 Fires in Brazil. (Source: Manaus AM, Brazil {CC BY 2.0} via Wikimedia Commons)

Fig. 12.2 Slash-and-burn practices are commonly used to convert rainforest to agricultural uses. (Source: Matt Zimmerman (CC BY 2.0) via Flickr)

While the focus of this unit will be on climate change and the Brazilian rainforest, it is important to consider other threats facing the forest:

- *Cattle ranching*: With Brazil as the world's top exporter of beef historically, cattle ranching has accounted for more than three quarters of the Amazon's deforestation . Pasturing cattle in the Amazon is inefficient, with each head of beef requiring about 1 ha of land. Once an area has been grazed off, a new area must be deforested to continue to support the cattle. Current estimates are that beef and soy production to feed cattle are driving accelerating rates of deforestation in the Amazon.

- *Industrial logging*: Fueled by the high demand for a range of timber products, illegal logging is common in the Amazon. Use of forged permits, cutting more than authorized quotas, and stealing from protected areas are rarely monitored or prosecuted. Even legally permitted logging provides few incentives to replant trees, harvest efficiently, or sustainably manage long-term timber stocks. According to satellite data (Ennes 2021), industrial-scale logging affects almost 500,000 ha per year (Fig. 12.3).

- *Subsistence farming*: It is estimated that one third of tropical deforestation is caused at the local level by people who need the resources provided by the forest for their survival. Subsistence farmers in the Amazon basin often

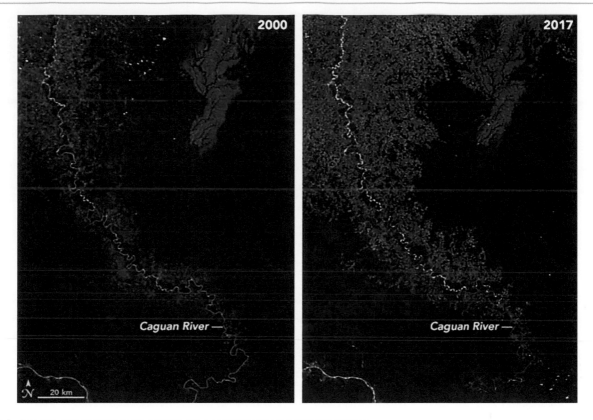

Fig. 12.3 Amazon deforestation from 2000 to 2017. (Source: NASA (Public Domain))

have relied on a slash-and-burn approach for clearing land for crops. While this approach has been used by indigenous people for centuries in a sustainable, small-scale rotational system, so many people currently depend on these forest resources that rotational cycles are shorter with little time for the forest to recover.

- *Agribusiness*. Large swaths of forest have been cleared for growing soybeans, starting oil palm plantations, and similar large-scale agribusiness efforts in the Amazon. Often, land deforested for grazing cattle is converted into soybean fields. Total area cultivated for soybeans alone has increased over the past two decades, from 600,000 ha in 1995 to 5 million ha in 2021. Ironically, the loss of forest driven by large-scale agriculture is actually creating conditions unsuitable for crop cultivation.

- *Gold mining*. Gravel deposits throughout the river channels and floodplains of the Amazon basin contain gold. In addition to removing forests to mine these deposits, secondary structures like roads and other infrastructure have added to the destruction of the forest. An additional threat is the toxicity of mercury required to extract the gold, polluting surface waters and contaminating fish that local villagers depend on.

Each of these activities typically includes some form of slash-and-burn activity. However, increased frequency and intensity of drought, reduced humidity due to forest loss, and increased human activity have resulted in a significant fire threat across the Amazon (Fig. 12.4). Unlike fire-adapted forests in the United States, fire is not a natural phenomenon in the Amazon, where nearly all fires are human-ignited.

Politics also likely play a role, with recent budget cuts and anti-environmental government leadership affecting efforts to protect Brazil's portion of the rainforest. Impacts of fires in the Amazon are far reaching, with significant GHG emissions and soot and black carbon from the fires carried in the atmosphere to glaciers in the Andes, setting off another climate-related positive feedback loop.

12.2.2 The Role of Climate Change

Connections between the Amazon rainforest and climate are strong, including how the Amazon impacts global climate and how climate impacts the Amazon. Intact forests sequester and store significant amounts of CO_2 historically absorbing almost a quarter of all GHG emissions. However, recent studies indicate that the Amazon rainforest is now emitting more CO_2 than it is able to absorb. Satellite imagery confirms that over the past decade, the Brazilian Amazon released nearly 20% more CO_2 than it absorbed. Evidence suggests that even intact forests are becoming net carbon emitters as

Fig. 12.4 Active fires detected across the Amazon during the 1-week period in August 2019. (Source NASA {Public Domain})

drought and heatwaves cause tree death, which further limits transpiration and the formation of rain clouds. This conversion of one of the world's largest carbon sinks to a net carbon source could have significant implications for the rate and extent of climate change.

While much of the attention in the Amazon rainforest is focused on CO_2, other GHGs are in play as well. Kimbrough (2021) noted that when trees burn, they release methane, a GHG more powerful than CO_2. Climate-associated warming of soils and sediments in the Amazon's seasonally flooded forests will release additional amounts of methane and nitrous oxide, another potent GHG.

In turn, increasing temperatures, drought, and shifting rainfall patterns are impacting the Amazon. Of particular concern is the link between ocean temperature and drought which results in cascading effects on the region's ecosystem, killing trees and leaving forests more vulnerable to fire. The result is a continuous feedback loop of changing climate conditions, impacted forests, and further changes in climate.

Models suggest that by 2050, Amazon temperatures may be 2–3 °C higher. Such increased temperatures and the altered rain patterns that will accompany them will likely further affect the region's forests, water availability, biodiversity, agriculture, and human health.

12.2.3 Solutions

The importance of the Amazon rainforest to global wellbeing has long been recognized, and a number of approaches are being used to try to preserve and restore the rainforest. While each of the solutions below has the potential to protect critical expanses of the Amazon rainforest, its fate has become a political football, with the leaders of South American nations showing widely varying levels of commitment to its protection.

Sustainable Land Management   A variety of approaches can be used to provide a source of income for residents without long-term impacts on the forest. Agroforestry (Fig. 12.5) mixes cultivated crops among the tree species, while polycultural management creates a patchwork of perennial crops, annual crops, pastureland, secondary growth, and forest. Olhos D'Agua ("Tears in the Eyes"), an agroforestry operation of 350 ha, features a complex system of trees and crop species, including high-quality cocoa beans. Other sustainable land management approaches focus on ecotourism, in which wealthy tourists pay substantial sums to stay in "green" resorts in the rainforest.

Carbon Offsets Carbon emitters in developed nations can invest in programs that protect existing forests or replant previously deforested areas and claim carbon credits to offset their CO_2 emissions. In the typical voluntary carbon market, each credit corresponding to one metric ton of removed CO_2 can be used to compensate for the release of a ton of GHGs elsewhere. In a related program, REDD+ (Reducing Emissions from Deforestation and Forest Degradation) offers incentives for countries to protect their valuable forest resources using sustainable management approaches.

Sustainable Silviculture Techniques While conventional harvesting typically focuses on tree extraction based on their marketable values, selective logging and reduced impact logging (RIL) can minimize impacts on the forest canopy, residual vegetation, and soil. Successful application of these techniques can cut waste from logging by up to 50% and allow for a rapid return to baseline forest function (Miller et al. 2011).

Community Based Protection Community based efforts are often effective at protecting ecosystems. In the Amazon, the Voluntary Environmental Agents Program (VEA Program) has empowered local communities in the Brazilian Amazon for the past 25 years. With its legal and formal regulatory structure, the VEA Program runs territorial surveillance and monitoring programs throughout the Amazon rainforest. The program has been replicated in 37 protected areas across the central Amazon, building on the knowledge and passion of local communities.

Soy Moratorium Implemented in 2006, the Amazon Soy Moratorium (ASM), in which international soy traders agreed to a ban on the purchase of soybeans from lands deforested after 2008, has been maintained by global commodities traders. The effort has been judged as successful, with fewer than 2% of the soy-growing areas out of compliance with the ASM.

While a host of possible management actions have been identified, to some extent, success in preserving the rainforest will only come if several remaining major hurdles are overcome. The Amazon basin is spread among nine nations with different political regimes and regulatory approaches. The basin is home to both an amazingly diverse animal and plant population but also millions of residents and hundreds of indigenous tribes that rely on the rainforest for their survival. One suspects that only the best planning by all the experts and an implementation plan embraced by all the political entities, local communities, and economic markets can succeed in protecting the vital role played by the Amazon rainforest.

12.2.4 Unit Challenge

"The lungs of the world" are in trouble. The Amazon rainforest is under siege, the victim of growing population pressures, global consumption demands, resource profiteering, and mismanagement. Despite decades of efforts by many individuals and groups to protect the Amazon rainforest, the battle is far from over. Assume that a very wealthy philan-

Fig. 12.5 Agroforestry system in initial phase with black pepper, **Piper nigrum**, as principal cash crop inter-planted with cupuacu, a unique Brazilian fruit. (Photo: R.C. Pinho et al. (CC BY SA 4.0))

thropist recently expressed interest in investing a large sum of money in efforts to protect the Amazon rainforest.

She is overwhelmed by the many different approaches that have been or are being used to manage this valuable global resource. She has come to you in your role as an environmental scientist known for your work studying sustainable management approaches in tropical forest systems to help her decide how to invest her funds. **Your Unit Challenge is to identify an investment approach that will not only protect large swaths of threatened rainforest in the face of ongoing climate change but will also support local communities.**

12.2.5 The Scenario

After reviewing the literature about possible approaches to sustainable forest management and conservation efforts in the central Amazon, you decide to analyze three possible solutions:

1. **Carbon credits**. Currently, global emitters of CO_2 can purchase carbon "credits" to offset their emissions to meet regulatory and policy requirements. Funds from the credits they purchase are used to invest in efforts to sequester and store carbon at other locations. Carbon credits were established for major corporations; fewer systems are available to allow individuals to offset some of the carbon footprint associated with their own consumption habits.
   Specific recommendation: Develop a new carbon market that would be available for individuals to purchase carbon credits to offset their personal consumption-based emissions. Your investment would help develop the technical infrastructure and on-the-ground support for a large-scale carbon credit program to be used specifically for reforestation efforts by local landowners across the Amazon.
2. **Silvopasturing.** By integrating trees, forage, and the grazing of domesticated animals, silvopasturing can support cattle farming while also sequestering carbon (Fig. 12.6). It is one of the oldest forms of agriculture and is particularly well suited for adoption in the Amazon rainforest.
   Specific recommendation: Support farmers throughout the Amazon rainforest who plant rapidly growing tree species known to be efficient at carbon absorption on their pasturelands. Funds would be used to purchase and maintain trees and pay stipends based on how much carbon was sequestered by converting grass-only pasture to silvopasture.
3. **Ecotourism.** The Amazon rainforest offers much to the globe-trotting wealthy tourist. Home to an incredible variety of plant and animal life, the Amazon already hosts a number of "green" tourist sites that put tourists in direct proximity to the rainforest and its diverse habitats. Ecotourism businesses support conservation of intact forests and local economies and help raise awareness.
   Specific recommendation: Develop a network of ecotourism travel destinations that would draw tourists to the central Amazon. Investments would support the purchase of the land, construction of lodging facilities, and operation costs. In addition to employing locals at a fair wage and serving as a sustainable development example, income from the $1000 per day resort fees would be used to purchase and preserve surrounding rainforest with the goal of building a connected network of protected forest.

12.2.6 Relevant Facts and Assumptions

Carbon credits

- A market analysis estimates that there is demand for approximately 100,000 metric tons of carbon credits per year.
- One ha of tropical rainforest stores about 250 metric tons of carbon annually. Large trees capture carbon more efficiently than smaller trees. Assume a carbon price of $50 US per metric ton.
- Restoring deforested lands in the Amazon rainforest costs an estimated $1140 per ha with planting densities ~5000 seedlings per ha.
- Start-up costs for carbon market development and marketing will be ~$1 million, with an additional $500,000 per year for carbon market management and on-the-ground monitoring of implementation.

Silvopasture

- A market analysis estimates that there are approximately 2000 ha of currently grass-only pasture for cattle that could be converted to silvopasture across the Amazon basin.
- One ha of silvopasture stores about 65 metric tons of carbon annually.
- Converting grass-only pasture to silvopasture with ~40% tree cover costs an estimated $750 per ha.
- Cattle ranchers will need transition incentives and supplemental support to offset initial losses while herds are rotated to ensure establishment of seedlings. Assume that you will need to offer $500/ha/year for the first 10 years of the program.
- Outreach and monitoring costs approximately $250,000 in the first year and $100,000 in each subsequent year.

Fig. 12.6 Livestock grazing in silvopasture paddock. (Source: National Agroforestry Center {CC BY 2.0} via Wikimedia Commons)

Ecotourism

- A market analysis estimates that there is sufficient demand to support three 100-ha eco resorts with 35-unit capacity.
- One ha of forested eco resort stores about 200 metric tons of carbon annually.
- 100 ha of Amazon rainforest can be purchased for $175,000 US to create one eco resort. Developing an ultra-low-impact 35-unit eco resort on this land will cost $800,000 US, with annual operational costs of around $150,000/year.
- Units would rent for an average of $1000 per night. Assume that year-round occupancy for each 35-unit site is 75%.
- Any revenue would be reinvested to help support operation of the resorts.

Each of these three approaches has important advantages and disadvantages and costs and benefits. Your task is to evaluate each and recommend the one that will, in your opinion, provide the maximum protection and restoration of the rainforest over time, while minimizing costs and supporting local communities.

12.2.7 Build Your Foundational Knowledge

Below are web sources that provide additional information about each of the solutions you're considering for this Unit Challenge. This information provides a critical foundation to help you evaluate each option and support your final choice. After reviewing each source, be prepared to answer questions in the Preparation Assessment Quiz and to summarize any information relevant to your Unit Challenge.

Carbon markets:
Can Markets Save the Amazon Rainforest?,
Best carbon Offsets for Individuals

Silvopasture:
Six Key Principles for a Successful Silvopasture,
How cattle can save the Amazon

Ecotourism:
Could ecotourism help protect the Amazon rainforest?,
Sustainable Tourism and the Amazonian People

Final Product: A one-page Fact Sheet summarizing your proposed investment strategy to conserve large swaths of the Amazon rainforest while supporting local communities and economies. Be sure to demonstrate how your proposed solution will stand up to the challenges posed by climate change.

12.2.8 Preparation Assessment Quiz

Are you ready to tackle your challenge? At this point you should understand the basic environmental principles and ecological processes involved in this environmental problem. Consider the following questions. If you are comfortable with answering these, then you are ready to head into Discovery, Analysis, and Solutions activities.

- Why is the Amazon rainforest so important in the fight to manage climate change?
- The Amazon rainforest is important for a number of other reasons. Discuss any two of these.
- What are primary threats to the forested ecosystems in the Amazon?
- What are the benefits of managing livestock using silvopastural systems?
- How do carbon markets function and what role might they play in conserving the Amazon forests?
- According to Smoot's article on carbon offsets for individuals, what are the four criteria for any meaningful offset program?
- According to "How Cattle Can Save the Amazon," how much carbon is released annually by clearing forests in the Amazon for cattle?
- According to the article in Business Destinations, what are the two key priorities for promoting sustainable ecotourism in the Amazon rainforest?
- For each of the proposed solutions, are there any additional benefits that might arise that might not be directly related to the Unit Challenge?
- For each of the proposed solutions, are there any negative unintended consequences that might result from their implementation?
- What additional information did you glean from your web sources that might help inform your Unit Challenge?

12.3 Protecting the Amazon: Discovery

Specific Skills You'll Need to Review: Navigating the Scientific Literature, Science Communication, Problem-Solving

12.3.1 Independent Research

(Key Skill: Navigating the Scientific Literature)

To better understand the potential for each of your approaches for protecting the Amazon rainforest, you first need to examine the literature to see what others have found. Conduct a search of the peer-reviewed scientific literature focused on the solution you have been assigned, and identify one research paper that focuses on your assigned approach.

Prepare a summary of the article you selected that includes the following:

- **Citation**
- **Main topic**: Stick to a few words, likely pulled from the title.
- **General summary**: A few bulleted sentences summarizing the research question it addresses and approaches it takes.
- **Methods:** How did they approach their research question?
- **Location**: Where was the work done?
- **Conclusions**: Concise list of the findings, specifically capturing the take-home message.
- **Relevance**: How might this study help inform your Unit Challenge? Feel free to make a bulleted list of information you may want to include later.

12.3.2 Literature Share: Reciprocal Instruction

(Key Skill: Scientific Communication)

Share In small groups, share and critique the research article you found. Keep in mind that your peers have not read this article, and it is your job to convey the key information to them. Note the items that will be important to consider when you are developing your solution to the challenge.

Critique Evaluate how these studies might help inform your Unit Challenge. Consider the following:

- Source (Quality of the work or bias of the authors)
- Methods (Did their methods sufficiently address the research question?)
- Conclusions (Did the results justify the conclusions made?)
- Relevance (Can these findings be applied to your challenge?)

Based on your critique, choose one article to share with the larger class, along with the key information that may be useful in deciding on a solution to propose.

12.3.3 Think-Compare-Share

(Key Skill: Problem-Solving)

Now that you have more information about possible solutions for this unit's challenge, you need to **develop a more formal problem definition** to guide your work throughout the rest of the exercises.

Think Start by working independently to develop a specific Problem Statement to guide the remainder of your work. Problem Statements provide the relevant information and boundaries to make the issue something you can effectively assess and tackle. The basics of a formal Problem Statement include the following:

Problem Statement: A short, concise statement summarizing the issue that includes the following:

- A **description** of the undesired condition or change that you hope to achieve (What is the actual problem?)
- **Justification** for addressing the problem (Why does this problem matter?)
- Potential **sources** or **causes** of this problem (What is the cause you need to address?)
- The **metrics** you will use to assess the status of the problem (How will you know if you are making a difference in the problem?)
- The **desired outcome** for these metrics (What is the end goal or ideal state?)
- Potential **solutions** to consider (How might you attempt to achieve this goal?)

Compare/Share Now return to your small group to share your Problem Statements. Use each of your ideas to develop a joint Problem Statement that contains all key information and is concise, clear, and well written.

Unit Discovery Summary *Submit a final Problem Statement that succinctly captures the key information to guide your work on this Unit Challenge.*

12.3.4 Reflecting on Your Work

(Key Skill: Personal Reflection)

After your work in Discovery, you should have a better idea of the problems you face and have produced a Problem Statement you can use to tackle the Unit Challenge. Take a moment to reflect on this work. Consider the following prompts but feel free to expand on any to best capture your learning experience and better inform your next steps.

- Of the skills you practiced in Discovery, which were the most challenging? Which were the most interesting?
- How were you most comfortable working during these exercises? In small groups, independently, or with the larger class? Why? How does your choice reflect your personality type and leadership style?

- Was your Problem Statement strictly focused on the problem of protecting the Amazon rainforest, or did it also consider important social and economic considerations? How might a focus on the environmental aspects limit your ability to identify truly sustainable solutions?
- You've been given three viable solutions to assess as a part of this case study. But this is not an exhaustive list of options or even necessarily the best possible course of action for every scenario. Take a moment to "think outside the box." Are there any other possible solutions you think would be worth exploring? Describe one that you think would be worth pursuing.

12.4 Protecting the Amazon: Analysis

Specific Skills You'll Need to Review: Quantitative Literacy, Sustainability Science

Review your Background and Discovery sections before beginning the Rotating Station exercises below. While you focused on one potential solution in your Independent Research in Discovery, keep an open mind as your work through Analysis activities.

12.4.1 Rotating Stations

(Key Skill: Quantitative Literacy)

At each of the following stations, you will review data that are relevant to the three potential solutions you're considering. Spend some time working through the analyses at each station to learn more about this issue and possible solutions for your Unit Challenge.

Be sure to write down one finding at each station that will help inform your selection of a solution.

Station 1: Carbon Credits Carbon credits are employed to avoid GHG emissions or to remove CO_2 from the atmosphere. However, these credits are estimates based on regional averages of forest stand dynamics and site conditions. Over-crediting occurs when these regional estimates calculate reimbursement for more carbon credits than are actually produced, and undercrediting occurs when forest stands produce more carbon per acre than regional estimates calculate for carbon credits.

Badgley et al. (2021) studied carbon offset projects in California to evaluate whether the carbon credits paid for were fully offset by carbon sequestration associated with their forest management actions. Such analyses allow policymakers to evaluate whether carbon offset credits paid for will result in actual climate benefits.

In the figure below, green dots represent the difference in carbon credits paid out using the standard common practice algorithm vs a more ecologically robust algorithm that better represents actual carbon sequestration in these stands. When the standard common practice algorithm is higher than the more ecologically robust algorithm (a net positive value in Fig. 12.7), more carbon credits were awarded than merited based on the amount sequestered. This is referred to as over-crediting.

Based on Badgley et al.'s analysis, answer the following about carbon offset projects in California:

- On average, how do the values calculated using the regional common practice algorithm compare to the actual carbon sequestered on these projects? What are the implications for the carbon credit market based on these findings?
- What might account for the variability in percent crediting error across projects?
- Common carbon credit calculations will undervalue, and hence undercompensate, landowners for carbon seques-

tering management practices on very rich, productive forest sites, while overvaluing, and hence overcompensating, landowners for carbon sequestering management practices on relatively poor, unproductive forest sites. How does this influence where landowners seek carbon credits and how might it incentivize landowners to continue with business-as-usual harvesting?

Record any relevant Station 1 findings

Station 2: Silvopasture Planting trees in pastures has the potential to increase the sustainability of cattle production in the Amazon by maintaining forage for cattle while sequestering carbon and maintaining biodiversity. Dablin et al. (2021) measured the potential for three different tree species to provide tree forage to pasture-fed cattle while building overall biomass stocks at a trial farm in Peru (Fig. 12.8).

Based on the data in the figure above, answer the following about how planting trees in pastures may impact both forage available for cattle to eat and overall biomass storage in the nonedible biomass accumulated:

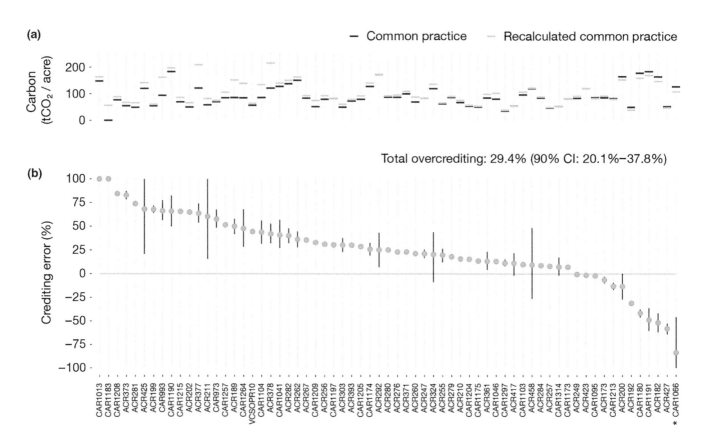

Fig. 12.7 Estimated carbon crediting error by project (x-axis). The authors compared the common regional calculations for carbon credits to a recalculated value based on a more ecologically robust calculation. Green dots represent the median estimate of over and undercrediting as a percentage of actual credits awarded for each project, with vertical black lines spanning the 25th and 75th percentile estimates. (Source: Badgley et al. 2021 (CC BY 4.0))

- How did the mean biomass of edible forage differ among the three tree species plantings and the grass-only control prior to grazing?
- Based on these data, did planting the tree species for silvopasturing impact the amount of edible forage available for cattle?
- What are other benefits of silvopasturing not captured in this figure of biomass accumulation?

Record any relevant Station 2 findings

Station 3: Ecotourism Tropical forests, like all ecosystems, provide a variety of ecosystem services, non-monetary benefits provided to humans. Brouwer et al. (2022) estimated the value of such services provided by the Brazilian Amazon. Calculations of the monetary value of ecosystem services include both use values (e.g., ecotourism, hunting and fishing) and non-use values (e.g., carbon regulation, water quality).

Use the information in the table to answer the following questions:

- Based on the estimated mean value of each ecosystem service in Table 12.1, which ecosystem services can be considered to have the highest "return" for conservation efforts?
- What might account for the large confidence intervals calculated for the various ecosystem services?
- How do you think the ecosystem services listed might be affected by climate change over the next decade? Which may be the most at risk of diminishing "returns"?
- Considering the mean ecosystem service value for ecotourism, how much in recreational and ecotourism returns might we expect for our proposed 100-ha resort?

- Assume that this 100-ha resort also provides carbon regulation, water cycle, and habitat-for-species ecosystem services. What is the total ecosystem service value for our 100-ha resorts?

Record any relevant Station 3 findings

Station 4 Your investor is focused on how to maximize the conservation of forest resources, while supporting local communities and economies, but she is also interested in maximizing the impact of her investment by preserving or rehabilitating as much rainforest area as possible. To help inform this financial aspect of the Unit Challenge, earlier you were given a set of Relevant Facts and Assumptions.

Now do some calculations to estimate how each solution achieves your goals of storing carbon while minimizing costs.

- Based on the market-demand estimates, calculate the year 1 start-up costs for each option as well at the 10-year operational costs.
- Calculate the total Mt C that could be stored using each of these treatments.
- Based on your calculations, which solution would be the cheapest? Which would store the most carbon? Which would be the most cost-effective (e.g., lowest cost per Mt C stored)?

Record any relevant Station 4 findings

12.4.2 System Mapping

(Key Skill: Sustainability Science)

Fig. 12.8 Mean average dry biomass of edible (available for forage) and non-edible (carbon storage) fractions of grass (Mg ha⁻¹) produced in four treatments: **Erythrina berteroana** (EB), **Inga edulis** (IE), **Leucaena leucocephala** (LL), and a no-tree, grass-only control (C). Destructive sampling was done prior to and post the introduction of cattle. (Source: Dablin et al. 2021 (Open Access))

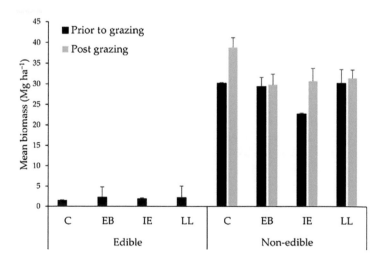

Table 12.1 Average economic values (2020 USD/ha/year) of selected ecosystem services for the Brazilian Amazon

	Mean value[1]	95% confidence interval	Min-max value	N[2]
Aggregate value for all ecosystem services	411.2 (122.7)	165.0–657.5	0.04–4,354.6	53
Ecosystem service				
Carbon regulation	333.3 (144.3)	−37.7–704.3	59.2–1032.7	6
Water cycle	150.8 (74.6)	−86.6–388.2	2.1–352.9	4
Recreation and ecotourism	410.3 (328.1)	−365.6–1186.3	0.06–2649.6	8
Habitat for species	454.6 (170.1)	109.0–800.2	0.04–4354.6	35

Source: Brouwer et al. 2022 (Open Access)
[1]standard error between brackets
[2]number of observations

Your assigned solution may have a variety of direct and indirect economic, social, and ecological impacts that should also be considered. For example, silvopasture techniques may still result in a loss of intact virgin forest lands (ecological), while ecotourism may provide employment and a source of income for local workers (economic), thus improving the lives of their families (social).

Working with other class members assigned the same solution, develop a simple sustainability map that shows the various system connections across each of the three sustainability domains (ecological, economic, and societal). When you find connections between impacts that cross domains, draw a line to identify the connection.

The goal is to think broadly about this larger system and envision how implementing actions in one domain (e.g., ecological) may impact components in another domain (e.g., economic or societal).

Use the following template (Fig. 12.9) to get started:

When each group has completed their basic sustainability map, **come together as a class to compare maps for the three possible solutions.** This information will help inform your decision support work later in the unit.

Unit Analysis Summary *Based on your explorations, what have you learned that can help inform your choice of a solution? Do the data support the adoption of one or more than one of these potential solutions?*

12.4.3 Reflecting on Your Work

(Key Skill: Personal Reflection)

During your work in Analysis, you explored some of the research into possible solutions to help inform your decision. Take a moment to reflect on this work. Consider the following prompts but feel free to expand on any to best capture your learning experience and better inform your next steps.

- How did you feel working with data? Do you consider quantitative literacy a strength or an area for improvement for you?

- How important should science be in informing management and policy? Do you feel the data you examined support and justify the costs of addressing your challenge?
- Any solution should be examined using a sustainability science lens. Your goal is to use this investment to maximize rainforest area conserved while supporting local communities. But any solution you implement may have other direct and indirect impacts. What are other possible economic, social, or ecological impacts? How should these considerations influence your decision?

12.5 Protecting the Amazon: Solutions

Specific Skills You'll Need to Review: Problem-Solving, Decision Support, Communicating Science

Review your Unit Challenge and the major findings from Discovery and Analysis, including the sustainability map you created to highlight connections among possible solutions and the larger socioecological system.

12.5.1 Small Group Guided Worksheets

(Key Skill: Decision Support)

Decision support matrices can help break down the desired outcomes to reflect multiple criteria for consideration, and they allow you to compare how each possible solution achieves those desired outcomes. This not only helps inform decision-making, it provides transparency in the decision process and justification to help you advocate for adopting a particular solution.

For the three possible solutions, you will evaluate how well each can achieve the following desired outcomes:

- Maximizes the area of rainforest that can be protected
- Promotes biodiversity over the long-term, given a changing climate
- Benefits local communities economically
- Is widely acceptable to stakeholder groups

Fig. 12.9 Template sustainability map to help connect ecological, societal, and economic considerations associated with your potential solutions

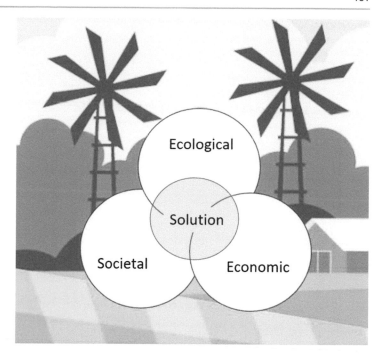

- Minimizes secondary impacts (e.g., economic, environmental)

Considering your three potential solutions, develop a formal decision support matrix to compare and evaluate each approach using the template matrix below (Table 12.2). Getting the most out of the decision matrix requires a depth of knowledge about each of the three possible solutions. Below we list additional sources about each option. Please review each of these, paying particular attention to the two alternatives to your assigned approach.

Note that your group may have uncovered a novel solution not included in this list of three. You may choose to work through the structured decision matrix with your self-identified solution as a fourth solution option.

12.5.2 Additional Sources Note: in Silvopasture line, separate "andhelp"

1. Carbon markets: How does carbon trading work? | World Economic Forum(weforum.org)
2. Silvopasture: Silvopasture could tackle Colombian Amazon's high deforestation rates andhelp achieve COP26 targets (phys.org)
3. Ecotourism: Climate Change and the rise of Ecotourism | by UN CC:Learn | Medium

Score each of your three potential solutions for each of the desired outcomes using a simple relative scale of 1 for least benefit to 5 for greatest benefit. Using a relative scale means you don't need to know exactly how well each solution meets the

goal of each desired outcome, but you can use your judgment to assess how well each solution works compared to the others.

While this is a relative (subjective) scale, note that you will need to justify your scoring of each solution for each desired outcome.

Once each cell in your decision matrix has a relative score, calculate an average score for each solution. Based on this analysis, which is the "best" solution, considering all your desired outcomes?

12.5.3 Role Playing

(Key Skill: Communicating Science)

In tackling environmental issues, you'll often find yourself working with groups of people with different perspectives about implementing a solution. You should use clear, concise communication to summarize your solution and justify its selection. To be effective, you must address concerns likely to be presented by various stakeholder groups and how the risks of taking these actions outweigh the risks of taking no action.

Your group will present and justify your chosen solution to the class with a particular emphasis on how it might benefit or impact one of the key stakeholder groups listed below. Be sure to include arguments that support this solution from the ecological, societal, and economic domains of sustainability science. Listeners will also be **assigned a stakeholder identity**, with an opportunity to **ask follow-up questions** after your presentation that reflect their unique concerns and perspectives. Your job will be to listen

Table 12.2 Basic decision matrix to compare the three possible solutions for your Unit Challenge. Include your justification for each solution's ranking for each desired outcome

Solution Options	Desired Outcomes				
	Maximizes area protected	Promotes biodiversity	Supports local economies	Minimizes secondary impacts	Maximizes stakeholder buy-in
Carbon credits					
Silvopasture					
Ecotourism					
Justification for Ranking					

Use a scale of 1-5 to score each option under each desired outcome category where 1 = does not achieve the desired outcome and 5 = completely achieves the desired outcome.

carefully and tailor your answers to this audience of stakeholders.

Key stakeholder groups include the following:

- Your primary investor
- Members of indigenous tribes living in the rainforest
- Local farmers and cattle ranchers
- Representatives of the Rainforest Alliance, an NGO dedicated to saving the rainforest
- Private sector interests wanting to invest in protecting the Amazon rainforest

12.5.4 Reflecting on Your Work

(Key Skill: Personal Reflection)

During your work in Solutions, you have explored several approaches that could be taken to support conservation efforts in the Amazon rainforest. Take a little more time now to reflect on your findings and the skills you practiced. Consider the following prompts but feel free to expand on any of them to best capture your learning experience and feelings about this issue.

- Reflect on your work through a sustainability science lens. Does your solution address ecological, economic, and societal considerations? Which considerations do you think should carry the most weight in the decision? Why?

- How did you weigh solutions that might have the greatest environmental impact against those that are most likely to be implemented and maintained for long-term impact?
- How can environmental scientists work to show the value of healthy ecosystems and justify the costs of mitigation strategies?
- Science communication can be challenging, especially when working with diverse audiences. We need to craft our communication to match the interests and values of the target audience, but how do you do this when your audience contains a mix of stakeholder groups? How can you maximize the impact of your message to a diverse audience?

Unit Solution Summary Summarize and justify your final solution choice and outline how it addresses the direct challenge while also considering social, economic, and ecological impacts. Also demonstrate that it will continue to meet the challenges posed by climate change.

12.6 Protecting the Amazon: Final Challenge

As a part of this Unit Challenge, you were asked to prepare a one-page Fact Sheet justifying an investment in various rainforest conservation activities. Your Fact Sheet should include the following components:

- Brief problem statement
- Recommended mitigation strategy with sufficient details to summarize the general approach
- Justification of this recommendation (e.g., long-term effectiveness, given anticipated climate changes, implementation costs, other benefits provided, etc.). Be sure to use a sustainability lens to include considerations of direct and indirect ecological, social, and economic considerations
- Any obstacles the group might face trying to implement your solution

Consider the use of figures, graphics, and tables to help summarize the system and how this solution is well suited to meet all desired outcomes.

Final Unit Challenge Submit your final recommendations in a one-page Fact Sheet using clear science communication designed for a lay audience.

References

Badgley G, Freeman J, Hamman JJ, Haya B, Trugman AT, Anderegg WR, Cullenward D (2021) Systematic over-crediting in California's forest carbon offsets program. Glob Chang Biol 28(4):1433–1445

Boulton CA, Lenton TM, Boers N (2022) Pronounced loss of Amazon rainforest resilience since the early 2000s. Nat Clim Chang 12:271–278. https://doi.org/10.1038/s41558-022-01287-8

Brandon K (2014) Ecosystem services from tropical forests: review of current science. Center for Global Development Working Paper (380).

Brouwer R, Pinto R, Dugstad A, Navrud S (2022) The economic value of the Brazilian Amazon rainforest ecosystem services: a meta-analysis of the Brazilian literature. PloS ONE 17(5):e0268425. https://doi.org/10.1371/journal.pone.0268425

Dablin L, Lewis SL, Milliken W, Monro A, Lee MA (2021) Browse from three tree legumes increases forage production for cattle in a silvopastoral system in the southwest Amazon. Animals 11(12):3585

Ennes JE (2021) Illegal logging reaches Amazon's untouched core, "terrifying" research shows. Mongabay.. https://news.mongabay.com/2021/09/illegal-logging-reaches-amazonsuntouched-core-terrifying-research-shows/

Kimbrough L (2021) We have turned the Amazon into a net greenhouse gas emitter: Study. Mongabay Series: Amazon Conservation. (mongabay.com)

Miller SD, Goulden ML, Hutyra LR, Keller M, Saleska SR, Wofsy SC, Figueira AMS, da Rocha HR, Camargo PB (2011) Reduced impact logging minimally alters tropical rainforest carbon and energy exchange. Proc Natl Acad Sci 108(48):19431–19435

Core Knowledge
Consumption, Solid waste management, Municipal planning, Life cycle assessments

13.1 Environmental Issue

The disposal of municipal solid waste (MSW) has long been associated with a host of environmental problems, including water and air pollution, habitat destruction, and aesthetic issues, among others (Fig. 13.1). In many developing nations, disposal of waste also attracts disease vectors and poses additional human health risks.

Links between waste management and climate change have been less publicized than other types of impacts but are no less important. Release of greenhouse gases (GHGs) like carbon dioxide (CO_2) and methane (CH_4) during various steps in the disposal process can be significant.

Recycling and waste transport have additional costs and benefits that must be considered.

Considering that the average US citizen generates almost 2.3 kg of MSW each day, sustainable waste management must be carefully considered by communities to identify the most efficient and effective ways to dispose of their trash.

13.2 Background Information

13.2.1 The Problem

Every human generates some type of waste (Fig. 13.2). Municipal solid waste (MSW) refers to all waste generated by people in their homes and businesses that is typically collected and disposed of by municipalities. Globally, we produce more than two billion tons of MSW annually. Amounts of MSW generated are influenced by economic activity, consumption rates, and population size. Per capita waste generation varies widely, with residents in wealthier countries

generally producing more waste. While the global average daily per capita production of solid waste is 0.75 kg, high-income countries average closer to 4.5 kg. Waste generation is also influenced by inefficient use of materials, excessive packaging practices, and cultural values.

The amount of MSW produced is a large part of this environmental issue, but how this waste is treated also has significant environmental ramifications. How MSW is handled is influenced by the types of waste produced. While landfills and dumps can be used to dispose of almost any type of waste, only organics and paper products can be composted. Metals and glass cannot be incinerated.

Currently, landfills are the most common repository for MSW. The composition of waste varies widely across nations and over time (Fig. 13.2). The largest component of MSW globally is organic material, but only a fraction of that is currently composted. Similarly, much of the paper, plastic, glass, and metal that could be recycled still finds its way to landfills, dumps, and incinerators.

The World Bank estimates that at least a third of MSW is not managed in ways that protect the environment. Environmental hazards associated with MSW disposal include the following:

- **Groundwater pollution:** Unlined or leaking lined landfills can contaminate underlying aquifers as leachate, or "garbage juice," produced by the waste seeps down through soils and bedrock. Newer landfills have leachate collection systems to reduce this risk.
- **Human health:** Improper disposal of waste can lead to a variety of adverse health outcomes. In addition to groundwater pollution, landfills' other threats include disease-carrying insects and rats, odors, and dust. Air pollution from incinerators has been linked to breathing problems, birth defects, and cancer.
- **Hazardous wastes:** Improperly handled hazardous materials, such as acids, toxic chemicals, and biological waste, can threaten the health of those living around waste dis-

J. Pontius, A. McIntosh, *Environmental Problem Solving in an Age of Climate Change*, Springer Textbooks in Earth Sciences, Geography and Environment, https://doi.org/10.1007/978-3-031-48762-0_13

Fig. 13.1 Virtually all humans generate some type of solid waste. How we manage this waste has direct connections to climate change and ecosystem and human health. (Source: SuSanA Secretariat {CC BY 2.0} via Wikimedia Commons)

Fig. 13.2 Global waste production over time by waste type and treatment. (Source Chen et al. 2020 (Open Access))

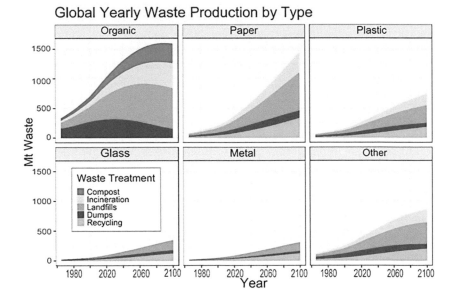

posal sites. Siting of such hazardous waste facilities in poor and underrepresented neighborhoods is a prime example of environmental racism/justice.

- **Ocean pollution:** Litter on beaches and illegal disposal of wastes from shipping both distract from aesthetics and threaten marine life. The Great Pacific Garbage Patch, which contains substantial amounts of plastic waste, now covers more than 1.6 million square km.
- **Greenhouse gas emissions (Fig. 13.3):** Transport of MSW from source to treatment facilities releases a variety of air pollutants. However, waste disposal makes additional direct contributions to GHG emissions. Currently, waste management accounts for approximately 5% of

total GHG emissions. This includes methane (CH_4) generation from landfills and a suite of GHGs emitted during trash incineration.

13.2.2 The Role of Climate Change

Connections between waste management and climate change may not be initially apparent, but how we create and dispose of trash is closely tied to carbon emissions and our overall "garbage footprint." It is possible to sort the effects of MSW disposal on climate change into direct and indirect impacts:

Fig. 13.3 Total reported emissions by waste subsector from the US EPA Greenhouse Gas Reporting Program. (Source: US EPA.gov {Public Domain})

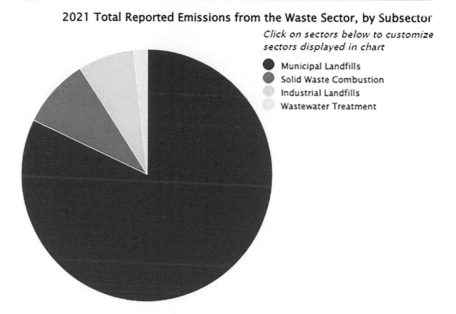

2021 Total Reported Emissions from the Waste Sector, by Subsector

Click on sectors below to customize sectors displayed in chart

- ● Municipal Landfills
- ● Solid Waste Combustion
- ○ Industrial Landfills
- Wastewater Treatment

Direct Impacts

- **Landfill GHG generation:** Organic components of MSW disposed of in landfills will release CH_4 as they decompose. While some landfills collect the gas for use as an energy source, in many cases, it escapes to the atmosphere.
- **Incineration:** Incineration of MSW can release several GHGs, including CO_2, CH_4, and N_2O.
- **Organic waste:** Animal manure, food waste, and other organic wastes release CO_2 and CH_4 to the atmosphere when they are burned or composted.

Indirect Impacts

- **Consumption:** Designing products for widespread consumption and rapid replacement has become the norm in many developed countries. This "throwaway culture" drives demand for the production of new replacement products, which release significant quantities of GHGs during their production and delivery.
- **Transportation:** Getting waste from the curb to its ultimate disposal destination involves transportation, which releases GHGs. How far materials need to be transported varies widely, but as local landfills fill up, long-haul waste transport has become a booming business.

In addition to the many varied impacts that MSW disposal has on climate change, the practice is, in turn, affected by climate change. Fei et al. (2021) illustrated the global extent of the impacts of climate change on waste disposal sites (Fig. 13.4) and described how climate change could impact MSW disposal, including damage to or shortened lifespan of

landfill infrastructure such as surface covers, bottom liners, and embankments. Since many disposal sites are located near oceans or estuaries, sea level rise and frequent strong storms could erode landfill foundations, resulting in contamination of marine waters.

13.2.3 Solutions

The old saying goes "reduce, reuse, and recycle." Any steps that can be taken to lower rates of consumption and the subsequent need to dispose of MSW will obviously reduce the release of GHGs and help slow climate change. Some suggested steps include the following:

Recycling Efforts While progress in reducing the amount of goods produced is particularly challenging in our consumer society, access to recycling programs has been growing. In the United States and other developed nations, more than a third of the waste generated is recovered through recycling or composting. In Sweden, almost 50% of their waste is recycled, with much of the remainder incinerated to generate energy. To reach this impressive recycling rate, Sweden gives discount vouchers to those who use recycling machines and ensures that all residential areas have access to recycling centers.

Other recent advances include the development of innovative approaches to recycle waste in consumer products and packaging. Terracycle works directly with manufacturers to help design their products to facilitate recycling, make their products and packages from more recycled materials, and reduce single-use consumption through reuse platforms.

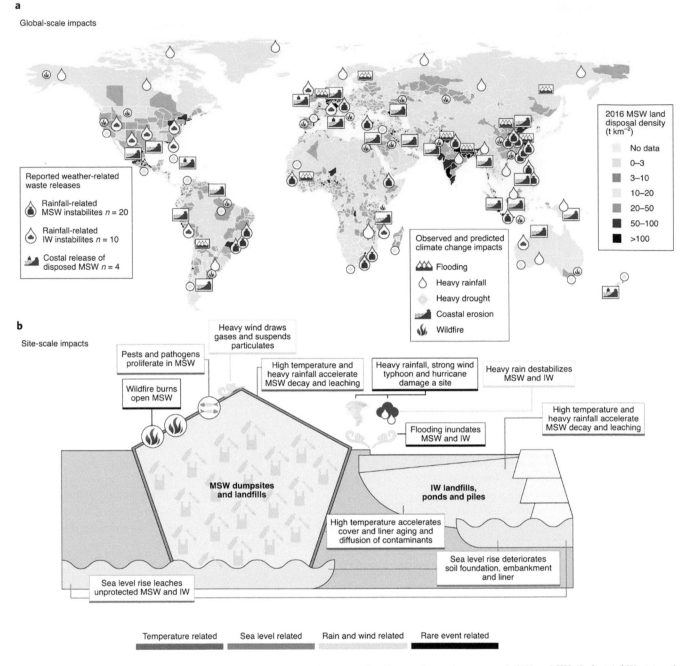

Fig. 13.4 (**a**) Global-scale MSW per-capita disposal density in 2016 overlaid with observed and predicted climate change impacts and reported weather-related waste releases with known locations. (**b**) Site-scale climate change impacts on MSW and IW *(Industrial Waste)* and the corresponding infrastructure. (Source: Fei et al. 2021)

The global waste recycling services market was valued at $55.1 billion US in 2020 but is expected to grow considerably in the coming years as consumer awareness about the environmental impacts of waste increases.

Reduced Consumption of Virgin Resources In addition to avoiding disposal in landfills or incinerators, recycling represents an important source of raw materials for manufacturing new products. Making goods from recycled materials typi-

cally requires less energy than making them from virgin materials. Less energy is needed to extract, transport, and process raw materials and to manufacture products when items are reused. With lower energy demand, fewer GHGs are emitted to the atmosphere.

For example, using recycled aluminum scraps to make aluminum cans requires 95% less energy than making the cans from bauxite ore, the raw material used to produce aluminum. Similarly, steel takes 75% less energy to make from recycled materials than from its raw material iron. A recent study by Razzaq et al. (2021) demonstrated that increased access to recycling will also lead to reduced GHG emissions. The authors found that every 1% increase in MSW recycling reduced overall carbon emissions by 0.32%.

Reuse Programs Historically thrift stores have existed to pass on household goods and textiles, but new programs are arising to facilitate reuse of many more consumer goods. The Repurpose Project was begun in 2011 to salvage items that traditional thrift stores didn't. By sorting, repairing, repurposing, and aggregating single-use items, project organizers find new uses for everything from office supplies to building materials.

Fig. 13.5 A fence made of glass bottles at Earthship Biotecture Visitor Center in Taos, New Mexico. (Source: Reettamarjaana [CC BY SA 4.0] via Wikimedia Commons)

Consumers Beyond Waste, an initiative of the World Economic Forum's Future of Consumption program, has developed a playbook for cities to develop and test reuse programs and partners with business, public, and governmental groups to facilitate new models of consumption. Others are finding novel ways to repurpose waste products, including using waste materials in building construction. Earthships, homes made from recycled materials including bottles (Fig. 13.5), tires, and cans use about one sixth the resources required for a traditional house.

Reducing the amount of MSW generated and disposed of in landfills and incinerators is ideal. But even when MSW is destined for disposal, there are new techniques designed to mitigate the potential impacts on the environment and reduce net GHG emissions.

Fig. 13.6 Methane gas is collected from this landfill near Glasgow Scotland where it can be compressed and stored for use as fuel. (Source: Roger May/Landfill / {CC BY-SA 2.0} via Wikimedia Commons)

Waste to Energy Burning refuse in incinerators to make electricity has been touted as key to reducing the net carbon emissions from waste treatment. However, the carbon intensity of incinerated waste is closer to coal and oil due to the increasing volume of carbon-based products in the waste stream. However, new technologies to ensure efficient energy recovery and sustainable disposal of ash waste can minimize the carbon footprint of incineration facilities.

Minimize Emissions from Landfills Organic materials dominate landfills, and their decomposition produces CH_4 gas. However, landfills can be tapped to collect this potent GHG and burn it to generate electricity or heat. Methane capture can achieve 85 percent efficiency or more in closed and engineered landfills (Fig. 13.6).

13.2.4 Unit Challenge

As your town's Environmental Officer, you are asked by the mayor to recommend the best way to manage the town's ever-increasing MSW generation. There is no organized recycling program in town, and the local landfill is at capacity and must be closed. You'd like to find a solution that will do the most to combat the effects of climate change, minimize other environmental and human health impacts, and make economic sense for the town.

13.2.5 The Scenario

After reviewing available data on the town's waste stream, you identify three possible solutions: (1) a new mixed recycling distribution center to be built in town; (2) a composting facility to be constructed on available land; and (3) a new secure landfill with a methane recapture system.

1. **Recycling distribution center:** The success of recycling depends on how easy it is for people to recycle. To increase access to recycling, facilities need to be nearby. While construction of recycling facilities to process various types of materials can be costly, distribution centers can be built for much lower costs. Distribution centers allow for recyclables to be collected and transported from local municipalities to larger recycling facilities where materials are processed.
    Specific recommendation: Build a recycling distribution center on current town land where residents can freely bring their recyclables for sorting and processing.
2. **Composting facility:** Building on the success of composting landscaping waste in many localities, there has been increased interest in composting more of the organic materials found in MSW. Successful programs require careful design to maximize biological decomposition and remove contaminants prior to reusing the resulting compost.
    Specific recommendation: Create a new composting facility on current town land where residents can freely bring their food waste and yard scraps for composting.
3. **Secure landfill:** Technology has advanced to the point where it is possible to isolate MSW from the environment. For example, lining landfills can prevent leachate from reaching groundwater systems. Leachate in the new secure landfill would be collected and treated at the town's wastewater treatment facility. Methane produced by the waste could also be used as an additional energy source.
    Specific recommendation: Develop a new secure landfill with methane recapture technologies on available town land.

13.2.6 Relevant Facts and Assumptions

- Currently, the town is paying $0.10 per kg to dispose of residential waste in a nearby landfill. This will need to be continued for any waste that can't be handled by the composting or recycling solutions.
- Each household in your town of 25,000 generates 50 kg of MSW annually that will need to be properly managed.
- Based on the proportion of various types of waste in this annual total, 55% could be composted; 67% could be recycled; and all could be disposed of in a new local landfill.
- Residents would likely compost 50% of their compostable waste and recycle 75% of their recyclable waste.
- Construction and operational costs for each of the solutions are as follows:
    - Recycling: $300,000 initial construction and $75,000 to operate yearly
    - Composting: $200,000 initial construction and $75,000 to operate yearly
    - Landfill: $450,000 initial construction and $100,000 to operate yearly
- Revenues generated from each of the solutions is as follows:
    - The sale of compost to replace chemical fertilizers at local farms nets approximately $0.25 per kg of initial compostable waste.
    - The sale of recyclables nets approximately $0.15 per kg of initial recyclable waste.
    - The sale of gas from landfills to produce energy nets approximately $0.05 per kg of total MSW.
- Carbon emissions reductions for each solution are as follow:

– Switching to recycling waste will save 1.1 Mt CO_2 equivalent emissions per kg of waste diverted from the landfill.

– Diverting organic waste to a composting facility will save 1.7 Mt CO_2 equivalent emissions per kg of waste diverted from the landfill.

– Capturing GHG from a secure landfill will save 0.33 Mt CO_2 equivalent in emissions per kg of waste moved to the landfill.

Each of these three approaches has important advantages and disadvantages and costs and benefits. Your task is to evaluate each and recommend the one that will, in your opinion, provide the best solution to manage the town's solid waste, while minimizing GHG emissions, minimizing secondary impacts, and garnering community support and buy-in.

13.2.7 Build Your Foundational Knowledge

Below are web sources that provide additional information about each of the solutions you're considering for this Unit Challenge. This information provides a critical foundation to help you evaluate each option and support your final choice. After reviewing each source, be prepared to answer questions in the Preparation Assessment Quiz and to summarize any information relevant to your Unit Challenge.

Recycling Distribution Centers:
EPA: Frequent Questions on Recycling, EPA: Improving Your Recycling Program

Municipal Composting:
Cornell Composting: Municipal Solid Waste Composting, State of Composting in the U.S.

Secure Landfills:
The Basics of Landfills https://www.epa.gov/landfills/basic-information-aboutlandfills
Landfill methane capture https://drawdown.org/solutions/landfill-methanecapture

Final Product: A one-page Fact Sheet summarizing your proposed municipal waste management strategy to meet the town's needs and minimize environmental impacts. Be sure to demonstrate how your proposed solution will stand up to the challenges posed by climate change.

13.2.8 Preparation Assessment Quiz

Are you ready to tackle your challenge? At this point you should understand the basic environmental principles and ecological processes involved in this environmental prob-

lem. Consider the following questions. If you are comfortable with answering these, then you are ready to head into Discovery, Analysis, and Solutions activities.

- What are the major components of Municipal Solid Waste (MSW)?
- What important environmental impacts result from the disposal of MSW?
- What are some steps that are being implemented to reduce impacts of MSW?
- How does MSW affect climate change and, in turn, how is MSW affected by climate change?
- The US EPA's web site on frequent questions on recycling claims that recycling one ton of aluminum cans conserves how many BTUs of energy?
- According to Biocycle, what are two key reasons why we need more composting?
- Project Drawdown estimates that in 2018, landfill methane accounted for how much of the total electricity generated globally?
- For each of the proposed solutions, are there any additional benefits that might arise from their implementation that might not be directly related to the Unit Challenge?
- For each of the proposed solutions, are there any negative unintended consequences that might result from their implementation?
- What additional information did you glean from your web sources that might help inform your Unit Challenge?

13.3 Sustainable Materials Management: Discovery

Specific Skills You'll Need to Review: Navigating the Scientific Literature, Science Communication, Problem-Solving

13.3.1 Independent Research

(Key Skill: Navigating the Scientific Literature)

To better understand the potential for each of your MSW management solutions, you first need to examine the literature to see what others have found. Conduct a search of the peer-reviewed scientific literature focused on the solution you have been assigned and identify one research paper that focuses on your assigned approach.

Prepare a summary of the article you selected that includes the following:

- **Citation**
- **Main topic**: Stick to a few words, likely pulled from the title.

- **General summary**: A few bulleted sentences summarizing the research question it addresses and approaches it takes.
- **Methods:** How did they approach their research question?
- **Location**: Where was the work done?
- **Conclusions**: Concise list of the findings, specifically capturing the take-home message.
- **Relevance**: How might this study help inform your Unit Challenge? Feel free to make a bulleted list of information you may want to include later.

13.3.2 Literature Share: Reciprocal Instruction

(Key Skill: Scientific Communication)

Share In small groups, share and critique the research article you found. Keep in mind that your peers have not read this article, and it is your job to convey the key information to them. Note the items that will be important to consider when you are developing your solution to the challenge.

Critique Evaluate how these studies might help inform your Unit Challenge. Consider the following:

- Source (Quality of the work or bias of the authors)
- Methods (Did their methods sufficiently address the research question?)
- Conclusions (Did the results justify the conclusions made?)
- Relevance (Can these findings be applied to your challenge?)

Based on your critique, choose one article to share with the larger class, along with the key information that may be useful in deciding on a solution to propose.

13.3.3 Think-Compare-Share

(Key Skill: Problem-Solving)

Now that you have more information about possible solutions for this unit's challenge, you need to **develop a more formal problem definition** to guide your work throughout the rest of the exercises.

Think Start by working independently to develop a specific Problem Statement to guide the remainder of your work. Problem Statements provide the relevant information and boundaries to make the issue something you can effectively

assess and tackle. The basics of a formal Problem Statement include the following:

Problem Statement: A short, concise statement summarizing the issue that includes the following:

- A **description** of the undesired condition or change that you hope to achieve (What is the actual problem?)
- **Justification** for addressing the problem (Why does this problem matter?)
- Potential **sources** or **causes** of this problem (What is the cause you need to address?)
- The **metrics** you will use to assess the status of the problem (How will you know if you are making a difference in the problem?)
- The **desired outcome** for these metrics (What is the end goal or ideal state?)
- Potential **solutions** to consider (How might you attempt to achieve this goal?)

Compare/Share Now return to your small group to share your Problem Statements. Use each of your ideas to develop a joint Problem Statement that contains all key information and is concise, clear, and well written.

Unit Discovery Summary Submit a final Problem Statement that succinctly captures the key information to guide your work on this Unit Challenge.

13.3.4 Reflecting on Your Work

(Key Skill: Personal Reflection)

After your work in Discovery, you should have a better idea of the problems you face and have produced a Problem Statement you can use to tackle the Unit Challenge. Take a moment to reflect on this work. Consider the following prompts but feel free to expand on any to best capture your learning experience and better inform your next steps.

- Of the skills you practiced in Discovery, which were the most challenging? Which were the most interesting?
- How were you most comfortable working during these exercises? In small groups, independently, or with the larger class? Why? How does your choice reflect your personality type and leadership style?
- Was your Problem Statement strictly focused on the environmental problem of sustainable waste management, or did it also consider important social and economic con-

siderations? How might a focus on the environmental aspects limit your ability to identify truly sustainable solutions?

- You've been given three viable solutions to assess as a part of this case study. But this is not an exhaustive list of options or even necessarily the best possible course of action for every scenario. Take a moment to "think outside the box." Are there any other possible solutions you think would be worth exploring? Describe one that you think would be worth pursuing.

13.4 Sustainable Waste Management: Analysis

Specific Skills You'll Need to Review: Quantitative Literacy, Sustainability Science

Review your Background and Discovery sections before beginning the Rotating Station exercises below. While you focused on one potential solution in your Independent Research in Discovery, keep an open mind as your work through Analysis activities.

13.4.1 Rotating Stations

(Key Skill: Quantitative Literacy)

At each of the following stations, you will review data that are relevant to the three potential solutions you're considering. Spend some time working through the analyses at each station to learn more about this issue and possible solutions for your Unit Challenge.

Be sure to write down one finding at each station that will help inform your selection of a solution.

Station 1: Municipal Recycling An important way that recycling can address the issue of climate change is by eliminating the GHG-intensive process of converting raw materials into finished products such as aluminum cans and plastics that then must be disposed of. Orlov et al. (2021) analyzed the net GHG emissions for incineration and landfills in the Murmansk region of Russia, as well as how GHG emissions would change if those materials were recycled (Fig. 13.7).

Based on Orlov et al.'s analysis, answer the following about net GHG emissions across MSW treatment types:

- Which waste types are the primary contributors to GHG emissions in this region?
- Which waste types have the greatest potential to reduce emissions if recycled?

- It's hard to imagine that GHG from a landfill could exceed emissions from an incinerator. What can explain the higher emissions from landfills?
- We don't think of food as being a recyclable material, but in this study, a significant amount of GHG emissions is included in the estimates for recycling landfill-destined food waste. How is food recycled and how much could this contribute to carbon savings for this municipality?
- Considering the information in the figure, which types of waste do you believe are most important to recycle?

Record any relevant Station 1 findings

Station 2: Greenhouse Gas Emissions Across Waste Management Types While every approach to waste management generates GHG emissions, it is possible to design an ideal waste management strategy that involves multiple streams of waste to minimize overall emissions. Which combination of incineration, composting, recycling, and landfill outlets works best for any municipality depends on the mix of waste types and accessibility to various outlets.

Kristanto and Koven (2019) examined four different management scenarios that divert municipal waste in Depok, Indonesia, in various proportions to incinerators, anaerobic digestion, waste treatment units (recycling), composting, and controlled landfills. In Fig. 13.8, each scenario includes various combinations of these waste management outlets. The panel on the left includes emissions savings resulting from electricity generation that is a part of the waste management process at the incinerator and anaerobic digestion facilities. The panel on the right does not include these electricity generation emissions savings.

Based on the data in this figure, answer the following about GHG emissions associated with various methods of municipal waste management:

- Scenario 1 represents business as usual for this municipality. Describe the various waste management streams currently used in Scenario 1.
- As shown in the panel on the left side of the figure, the authors are exploring three alternative waste management streams in Scenarios 2, 3, and 4. What are the primary changes the authors are suggesting in the municipality's approach to waste management in these three new scenarios?
- How does the generation of electricity alter net emissions for each of the scenarios?
- Which scenario has the lowest GHG emissions if electricity generation is not an option in accessible facilities?
- Organic waste streams can be diverted from landfills and incinerators to composting and anaerobic digestion facilities. While both have considerably lower GHG emissions

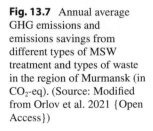

Fig. 13.7 Annual average GHG emissions and emissions savings from different types of MSW treatment and types of waste in the region of Murmansk (in CO_2-eq). (Source: Modified from Orlov et al. 2021 {Open Access})

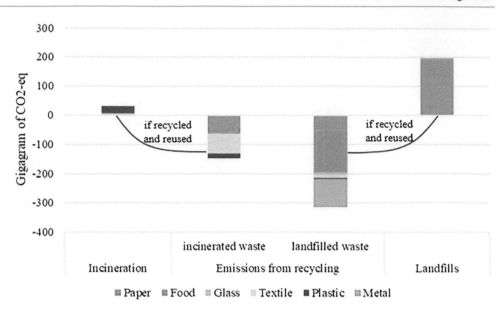

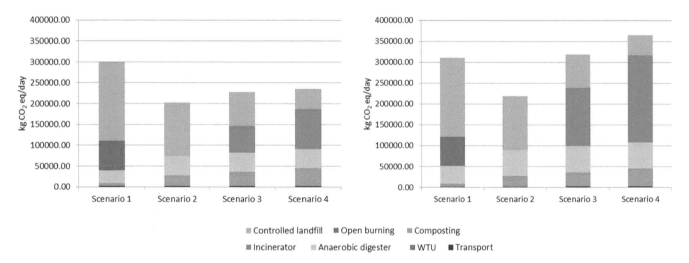

Fig. 13.8 The division of GHG emissions for four scenarios of waste management, with (left) and without (right) emissions savings from electricity generation. (Source: Kristanto and Koven 2019 {Open Access})

compared to incinerators and landfills, both waste management approaches also have other potential benefits. What are some other secondary benefits of managing organic waste via composting and anaerobic digestion?

Record any relevant Station 2 findings

Station 3: Secure Landfills An important process to consider when evaluating the impacts of landfilling MSW on the environment and climate change is the generation of GHGs. New technologies allow us to tap landfills to collect these gases for energy generation, but the amount of gases produced and captured varies over time and depends on environmental conditions and the type of waste in the landfill. These gases are produced in landfills whether or not they are collected and used for energy production.

Methane combustion for energy production offsets carbon emissions that would have been produced by other forms of electricity generation. Lee et al. (2017) examined the production of GHGs and the efficiency of their collection as part of a life cycle analysis of waste-to-energy pathways. In Fig. 13.9, they show the amount of CO_2 emissions from waste decomposition (yellow) compared to the amount of methane (CO_2 equivalent) that could be collected and combusted (blue) and the amount of methane (CO_2 equivalent) that remained in the landfill uncollected (gray) for four different types of waste.

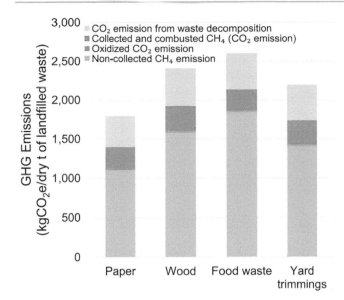

Fig. 13.9 GHG emissions from landfilled organic wastes for four types of waste materials (calculated as GWPs *(Global Warming Potentials)* of CO_2 and CH_4 over a 100-year time horizon). (Source: Lee et al. 2017 {Open Access})

Use the information in the figure to answer the following questions:

- How do total GHG emissions from landfills differ across the four types of organic waste? What might explain these differences?
- Assume that any GHGs not collected and combusted would simply be emitted to the atmosphere. Approximately what proportion of these total potential GHG emissions do the collection and combustion efforts capture?
- Consider that electricity generated by the combustion of GHG emissions from landfills leads to a displacement credit of 500 kg CO_2e/dry t of landfill waste based on the US power generation mix. How would these indirect downstream GHG savings alter your calculation of GHG emissions from collection and combustion efforts?

Record any relevant Station 3 findings

Station 4 A key consideration in your search for a solution is cost: which of the three options makes the most sense economically? Not only do the capital and operational costs of the three approaches differ, but there are some very different economic benefits associated with the three approaches. Landfills release methane, which can be burned for energy. Compost can be sold as an organic fertilizer to replace chemical fertilizers. There are also commercial markets for a number of recycled products. For each approach, review the data presented in the Relevant Facts and Assumptions section, and calculate the following:

- The net cost (total cost - revenue) to install and operate each solution over a 10-year period
- The total Mt CO_2 equivalent emissions savings for each solution
- The cost per Mt CO_2 equivalent savings over the 10-year period

Record any relevant Station 4 findings

13.4.2 System Mapping

(Key Skill: Sustainability Science)

Your assigned solution may have a variety of direct and indirect economic, social, and ecological impacts that should also be considered. For example, local residents may have different opinions about the development of a new composting center vs a recycling distribution center (social), with potential implications for home values in the vicinity of the facility (economic), while a landfill may have potential impacts on ground water (environmental).

Working with other class members assigned the same solution, develop a simple sustainability map that shows the various system connections across each of the three sustainability domains (ecological, economic, and societal). When you find connections between impacts that cross domains, draw a line to identify the connection.

The goal is to think broadly about this larger system and envision how implementing actions in one domain (e.g., ecological) may impact components in another domain (e.g., economic or societal).

Use the following template (Fig. 13.10) to get started:

When each group has completed their basic sustainability map, **come together as a class to compare maps for the three possible solutions.** This information will help inform your decision support work later in the unit.

Unit Analysis Summary Based on your explorations, what have you learned that can help inform your choice of a solution? Do the data support the adoption of one or more than one of these potential solutions?

13.4.3 Reflecting on Your Work

(Key Skill: Personal Reflection)

During your work in Analysis, you explored some of the research into possible solutions to help inform your decision. Take a moment to reflect on this work. Consider the following prompts but feel free to expand on any to best capture your learning experience and better inform your next steps.

Fig. 13.10 Template sustainability map to help connect ecological, societal, and economic considerations associated with your potential solutions

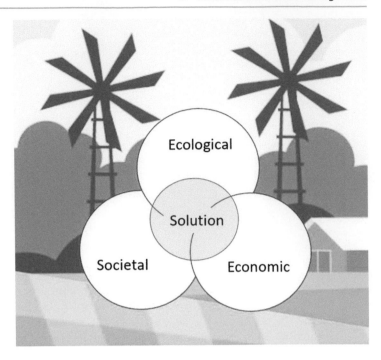

- How did you feel working with data? Do you consider quantitative literacy a strength or an area for improvement for you?
- How important should science be in informing management and policy? Do you feel the data you examined support and justify the costs of addressing your challenge?
- Any solution should be examined using a sustainability science lens. Your goal is to identify the best approach for managing the town's MSW while minimizing net GHG emissions. But any solution you implement may have other direct and indirect impacts. What are other possible economic, social, or ecological impacts? How should these considerations influence your decision?

13.5   Sustainable Materials Management: Solutions

Specific Skills You'll Need to Review: Problem-Solving, Decision Support, Communicating Science

Review your Unit Challenge and the major findings from Discovery and Analysis, including the sustainability map you created to highlight connections among possible solutions and the larger socioecological system.

13.5.1 Small Group Guided Worksheets

(Key Skill: Decision Support)

Decision support matrices can help break down the desired outcomes to reflect multiple criteria for consider-

ation, and they allow you to compare how each possible solution achieves those desired outcomes. This not only helps inform decision-making, but it provides transparency in the decision process and justification to help you advocate for adopting a particular solution.

For the three possible solutions, you will evaluate how well each can achieve the following desired outcomes:

- Reduces dependence on neighboring landfills to accept waste streams
- Minimizes GHG emissions from the town's MSW generation over the long-term, given a changing climate
- Provides economic benefits to the town
- Minimizes secondary impacts (e.g., economic, environmental)
- Is widely acceptable to town residents

Considering your three potential solutions, develop a formal decision support matrix to compare and evaluate each approach using the template matrix below (Table 13.1). Getting the most out of the decision matrix requires a depth of knowledge about each of the three possible solutions. Below we list additional sources about each option. Please review each of these, paying particular attention to the two alternatives to your assigned approach.

Note that your group may have uncovered a novel solution not included in this list of three. You may choose to work through the structured decision matrix with your self-identified solution as a fourth solution option.

Table 13.1 Basic decision matrix to compare the three possible solutions for Your Unit Challenge. Include your justification for each solution's ranking for each desired outcome

Solution Options	Desired Outcomes				
	Minimizes waste to neighboring landfills	Minimizes GHG emissions	Maximizes economic benefit	Minimizes secondary impacts	Maximizes stakeholder buy-in
Recycling distribution center					
Composting facility					
Landfill with GHG capture					
Justification for Ranking					

Use a scale of 1-5 to score each option under each desired outcome category where 1 = does not achieve the desired outcome and 5 = completely achieves the desired outcome.

13.5.2 Additional Sources

- Recycling distribution centers: Warehouse and Distribution Center Recycling - Generated Materials Recovery
- Composting facilities: https://www.epa.gov/sustainable-management-food/types-composting-and-understanding-process
- Landfillls with methane capture: https://www.epa.gov/lmop/basic-information-about-landfill-gas

Score each of your three potential solutions for each of the desired outcomes using a simple relative scale of 1 for least benefit to 5 for greatest benefit. Using a relative scale means you don't need to know exactly how well each solution meets the goal of each desired outcome, but you can use your judgment to assess how well each solution works compared to the others.

While this is a relative (subjective) scale, note that you will need to justify your scoring of each solution for each desired outcome.

Once each cell in your decision matrix has a relative score, calculate an average score for each solution. Based on this analysis, which is the "best" solution, considering all your desired outcomes?

13.5.3 Role Playing

(Key Skill: Communicating Science)

In tackling environmental issues, you'll often find yourself working with groups of people with different perspectives about implementing a solution. You should use clear, concise communication to summarize your solution and justify its selection. To be effective, you must address concerns likely to be presented by various stakeholder groups and how the risks of taking these actions outweigh the risks of taking no action.

Your group will present and justify your chosen solution to the class with a particular emphasis on how it might benefit or impact one of the key stakeholder groups listed below. Be sure to include arguments that support this solution from the ecological, societal, and economic domains of sustainability science. Listeners will also be **assigned a stakeholder identity**, with an opportunity to **ask follow-up questions** after your presentation that reflect their unique concerns and perspectives. Your job will be to listen carefully and tailor your answers to this audience of stakeholders.

Key stakeholder groups include the following:

- Residents served by the new waste management system
- Residents in the vicinity of a facility siting location
- Officials charged with MSW program operations

- Representatives of the Global Warming Coalition, an NGO dedicated to limiting CO_2 emissions
- Representatives of state and federal environmental agencies

13.5.4 Reflecting on Your Work

(Key Skill: Personal Reflection)

During your work in Solutions, you have explored several approaches that could be taken to more sustainably manage waste in your town. Take a little more time now to reflect on your findings and the skills you practiced. Consider the following prompts but feel free to expand on any of them to best capture your learning experience and feelings about this issue.

- Reflect on your work through a sustainability science lens. Does your solution address ecological, economic, and societal considerations? Which considerations do you think should carry the most weight in the decision? Why?
- How did you weigh solutions that might have the greatest environmental impact against those that are most likely to be implemented and maintained for long-term impact?
- How can environmental scientists work to show the value of healthy ecosystems and justify the costs of mitigation strategies?
- Science communication can be challenging, especially when working with diverse audiences. We need to craft our communication to match the interests and values of the target audience, but how do you do this when your audience contains a mix of stakeholder groups? How can you maximize the impact of your message to a diverse audience?

Unit Solution Summary Summarize and justify your final solution choice, and outline how it addresses the direct challenge while also considering social, economic, and ecological impacts. Also demonstrate that it will continue to meet the challenges posed by climate change.

13.6 Sustainable Waste Management: Final Challenge

As a part of this Unit Challenge, you were asked to prepare a one-page Fact Sheet justifying the implementation of a new management strategy for this town's MSW. Your Fact Sheet should include the following components:

- Brief problem statement
- Recommended waste management system with sufficient details to summarize the general approach
- Justification of this recommendation (e.g., long-term effectiveness, given anticipated climate changes, potential economic benefits, other benefits provided, etc.). Be sure to use a sustainability lens to include considerations of direct and indirect ecological, social, and economic considerations
- Any obstacles the group might face trying to implement your solution

Consider the use of figures, graphics, and tables to help summarize the system and how this solution is well suited to meet all desired outcomes.

Final Unit Challenge Submit your final recommendations in a one-page Fact Sheet using clear science communication designed for a lay audience.

References

Chen DMC, Bodirsky BL, Krueger T, Mishra A, Popp A (2020) The world's growing municipal solid waste: trends and impacts. Environ Res Lett 15(7):074021

Fei X, Fang M, Wang Y (2021) Climate change affects land-disposed waste. Nat Clim Change 11(12):1004–1005

Kristanto GA, Koven W (2019) Estimating greenhouse gas emissions from municipal solid waste management in Depok, Indonesia. City Environ Interact 4:100027

Lee U, Han J, Wang M (2017) Evaluation of landfill gas emissions from municipal solid waste landfills for the life-cycle analysis of waste-to-energy pathways. J Clean Prod 166:335–342

Orlov A, Klyuchnikova E, Korppoo A (2021) Economic and environmental benefits from municipal solid waste recycling in the Murmansk Region. Sustainability 13:10927

Razzaq A, Sharif A, Najmi A, Tseng ML, Lim MK (2021) Dynamic and causality interrelationships from municipal solid waste recycling to economic growth, carbon emissions and energy efficiency using a novel bootstrapping autoregressive distributed lag. Resour Conserv Recycl 166:105372

Index